SCUOLA NORMALE SUPERIORE

QUADERNI

Jerzy Zabczyk

Topics in
Stochastic Processes

PISA - 2004

ISBN: 88-7642-131-9

Jerzy Zabczyk
Institute of Mathematics
Polish Academy of Sciences
Śniadeckich 8, Office 512
00-950 Warsaw
Poland

Topics in Stochastic Processes

PREFACE

The notes are based on lectures on stochastic processes given at Scuola Normale Superiore in 1999 and 2000. Some new material was added and only selected, less standard results were presented. We did not include several applications to statistical mechanics and mathematical finance, covered in the lectures, as we hope to write part two of the notes devoted to applications of stochastic processes in modelling. The main theme of the notes are constructions of stochastic processes. We present different approaches to the existence question proposed by Kolmogorov, Wiener, Ito and Prohorov. Special attention is paid also to Lévy processes. The lectures are basically self-contained and rely only on elementary measure theory and functional analysis. They might be used for more advanced courses on stochastic processes.

In the introduction we motivate the subject and give some historical background. In the next chapter we gather basic definitions and results from measure theoretic foundations of probability theory. Special attention is paid to characteristic functions. Chapter 3 concerns Kolmogorov's existence theorem. We recall Carathéodory's extension theorem and prove a fundamental result of Ulam. After the proof of the Kolmogorov theorem we construct various classes of processes: Gaussian, Markov and Lévy as well as Wiener and Poisson.

Kolmogorov's existence result does not imply any properties of the trajectories of the constructed processes. This is why in Chapter 4 we present a method due to Doob of establishing existence of a stochastic process with right continuous trajectories having left limits. Processes of this type are called càdlàg processes, from French *continu à droite et pourvu de limites à gauche*. The method is based on martingale theory which is presented here in a self-contained way. We apply the method to prove existence of càdlàg versions of Lévy processes and Markov processes of Feller type.

Wiener's constructive approach is presented in Chapter 5. After general comments on expansions of stochastic processes we establish a Lévy-Ciesielski representation of a Wiener process as well as an explicit construction of a Poisson process.

Wiener and Poisson processes are the building blocks of the whole theory of stochastic processes. They are special cases of Lévy processes which are studied analytically in Chapter 6. We establish the so called Lévy-Khinchin formula exploiting the Hille-Yosida theorem of semigroup theory. Some basic results on the semigroups and generators corresponding to Lévy processes are presented as well, together with the subordination method.

Probabilistic representation of Lévy processes in terms of Poissonian measures is given in Chapter 7.

Chapter 8 explores a connection between Markov processes and stochastic equations. We first recall Courrège's analytic characterisation of Markov semigroups. Then we present Ito's construction of Markov processes as solutions of stochastic equations with Wiener and Poissonian stochastic terms.

Chapter 9 is devoted to stochastic integration in infinite dimensions and sets a basis for infinite dimensional generalisations of results from the previous chapter. We start from basic results on Gaussian measures on Hilbert spaces and introduce an infinite dimensional Wiener process. Then a stochastic integral with respect to an infinite dimensional Wiener process is constructed and the fundamental isometric formula is deduced. Next the stochastic integration with respect to square integrable, Hilbert valued martingales is presented. To prove a generalisation of the isometric formula two types of martingale brackets are introduced and a martingale covariance process is constructed.

The theory of weak convergence and tightness of measures on metric spaces is presented in Chapters 10 and 11. We give a detailed exposition of Prohorov's theorem in metric spaces. Some criteria of weak convergence and tightness in the space $C([0, T]; E)$ of continuous functions with values in a metric space E are developed in Chapter 10. The convergence of random walks to a Wiener process is given as application. We also introduce the factorisation method, due to L. Schwartz, to establish tightness and prove Donsker's invariance principle. Finally we prove Kolmogorov's continuity criteria in the context of weak convergence.

ACKNOWLEDGEMENTS. The idea to have a printed version of the notes is due to Prof. G. Da Prato and I thank him for this initiative. The notes were read and corrected by Prof. M. Fuhrman who also took the burden of typing the text. I am grateful to him for his help.

Jerzy Zabczyk

CONTENTS

CHAPTER 1

Introduction

Probability theory is concerned with random events or more generally with random variables. Initially, random variables were treated in an intuitive way as *variables which take values with some probabilities*. They were characterised by their distribution functions. If X is a random variable then its distribution function F is defined by the formula:

$$F(x) = \mathbb{P}(X \le x), \qquad x \in \mathbb{R}^1.$$

Thus $F(x)$ is the probability that X takes values not greater than x. Random functions or, as we say today, stochastic processes, were defined as families of random variables $X(t)$, $t \in [0, T]$. They were characterised by multivariate distribution functions

$$(1.0.1) \qquad F_{t_1, \dots, t_n}(x_1, \dots, x_n) = \mathbb{P}(X(t_1) \le x_1, \dots, X(t_n) \le x_n)$$

defined for all $0 \le t_1 < t_2 < \dots < t_n \le T$ and $x_1, \dots, x_n \in \mathbb{R}^1$. Mathematical treatment of stochastic processes was rather involved and the need to put the theory on firm mathematical foundations became apparent. This was done in 1933 by A. Kolmogorov [35]. Probability theory and the theory of stochastic processes became at that time a part of measure theory and thus a solid part of mathematics.

To see a need for a formalisation consider a model of the Brownian motion proposed by A. Einstein [22] in 1905. Let $X(t)$ be the projection of the position of a particle, suspended in a liquid, onto the first coordinate axis. It is a random variable in its intuitive sense and the behaviour of $X(t)$, $t \ge 0$, is very chaotic. The following postulates seem to be natural:

1) The process $X(t)$, $t \ge 0$, has independent increments in the sense that for any sequence $0 \le t_0 < t_1 < \dots < t_n$, $n = 1, 2, \dots$ and any numbers

$y_1, \ldots, y_n \in \mathbb{R}^1$:

$$\mathbb{P}(X(t_i) - X(t_{i-1}) \le y_i, \ i = 1, \ldots, n) = \prod_{i=1}^{n} \mathbb{P}(X(t_i) - X(t_{i-1}) \le y_i),$$

2) The trajectories $X(t)$, $t \ge 0$, should be continuous.

In fact the postulates are satisfied by many models and are not contradictory. If, however, we add the postulate:

3) The trajectories are differentiable,

then conditions 1)-3) are contradictory. More precisely, only a deterministic, constant speed movement satisfies all the conditions. This means that we cannot construct a realistic and non-contradictory model fulfilling 1)-3). It is therefore of great importance to deal with existence theorems and the notes are concerned with the results of that sort.

1.1. – Existence questions

Let us comment a bit more on the existence problem. Stochastic processes can be defined and characterised in various ways, for instance by some postulates like 1)-3) or by specifying their finite dimensional distributions (1.0.1). There are several ways of constructing stochastic processes. One, very general approach, introduced by A. Kolmogorov [35] in 1933, assumes that finite dimensional distribution functions $F_{t_1, t_2, \ldots, t_n}$ are given and satisfy some natural consistency conditions. The approach, called sometimes *Kolomogorov's approach*, consists in building a probability measure $\mathbb{P}$ on the space $\Omega = \mathbb{R}^{[0,T]}$ of all real functions defined on the time interval $[0, T]$. The measure $\mathbb{P}$ should be such that for the so called canonical process X:

$$X(t, \omega) = \omega(t), \qquad \omega \in \Omega, \ t \in [0, T],$$

all the identities (1.0.1) are true. For the probability that $X(t_1) \le x_1, \ldots, X(t_n) \le x_n$ we can now write a correct mathematical expression:

$$\mathbb{P}(\{\omega \ : \ X(t_1, \omega) \le x_1, \ldots, X(t_n, \omega) \le x_n\}).$$

A variation of Kolmogorov's method is due to Yu. V. Prohorov [52] and is based on the concept of weak convergence of measures on metric spaces. The laws of the processes considered are obtained as weak limits of the laws of directly given simple processes, see [3]. The so called martingale method of solving stochastic equations, due to D. Stroock and S. R. R. Varadhan [60], is of the same category.

A completely different method is attributed to N. Wiener. It is less general and starts from a specific countable or uncountable family of random variables and builds the required process from the elements of the family in a constructive manner. For the first time this approach was applied in a joint paper of N. Wiener and R. E. Paley [51] to construct a Brownian motion process. As the point of departure one takes a sequence of independent, normally distributed random variables ξ_n, $n = 1, 2, \ldots$, defined on $(\Omega, \mathcal{F}, \mathbb{P})$, with

$$\mathbb{P}(\xi_n \leq x) = \frac{1}{\sqrt{2\pi}} \int_{-\infty}^{x} e^{-\frac{y^2}{2}} dy, \qquad x \in \mathbb{R}^1, \, n = 1, 2, \ldots.$$

It turns out that a process W defined by the infinite series

$$W(t, \omega) = \sum_{n=1}^{+\infty} \xi_n(\omega) \, e_n(t), \qquad t \in [0, T],$$

with properly chosen deterministic functions e_n, has all the postulated properties.

A powerful version of the method was proposed by K. Ito [32]. Processes are constructed as solutions to stochastic equations which can be solved by iterative methods. To formulate the equations and investigate their properties K. Ito developed a powerful stochastic calculus.

1.2. – Some history

In the language of modern probability, a stochastic process is a family of random variables $(X(t), t \in \mathbf{T})$, where $\mathbf{T} \subset [0, +\infty)$, defined on a probability space $(\Omega, \mathcal{F}, \mathbb{P})$. Stochastic processes were invented to articulate and study random phenomena which evolve in time. First processes were introduced at the beginning of the twentieth century. In his 1900 doctoral dissertation L. Bachelier [1] regarded the movement of prices on stock exchange as stochastic processes and introduced, in an intuitive way, the so called Wiener process. Several years later in 1909 A. Erlang [26] was analysing the work of telephone exchanges, most modern devices of the time. He realised that the number of busy lines has a random character and depends on time in a capricious way. To solve the practical question of how many operators should work in an exchange he introduced a special case of what is now known as a Markov process with a finite number of states. In the same year F. Lundberg [44] published a paper in which he investigated the time evolution of the capital of an insurance company and used, for this purpose, compound Poisson processes. A Wiener process was constructed as a mathematical object in 1923 by N. Wiener [62] and the theory of stochastic processes, as a part of mathematics, was created by A. Kolmogorov in his 1933 paper on the axiomatics of probability theory [35]. The first monograph on the subject was published in 1952 by J. Doob [20]. Since then the theory of stochastic processes has been constantly growing and has become the most important part of probability theory and one of the most influential tools of applied mathematics, see [47], [7].

The mathematical theory of stochastic processes became possible due to a rapid development of measure theory in the first quarter of the twentieth century. We will list only the most important events. The concept of σ-field was introduced in 1898 by E. Borel [6]. Lebesgue's book on integration [42] appeared in 1902. Integration on abstract sets was initiated by M. Fréchet [28] in 1915. The fundamental extension theorem by Carathéodory [8] was published in 1918. The general version of the so called Radon-Nikodym theorem was proved by O. Nikodym [48] in 1930. The first monograph on the general measure theory was written by S. Saks [54] (Polish edition in 1930 and English edition in 1937).

CHAPTER 2

Measure theoretic preliminaries

We collect here basic definitions and results from measure theoretic foundations of the probability theory. Special attention is paid to characteristic functions.

2.1. – Measurable spaces

Let Ω be a set. A collection $\mathcal{F}$ of subsets of Ω is said to be a *σ-field* (*σ-algebra*) if

1) $\Omega \in \mathcal{F}$.
2) If $A \in \mathcal{F}$, then A^c, the complement of A, belongs to $\mathcal{F}$.
3) If $A_n \in \mathcal{F}$, $n = 1, 2, \ldots$, then $\bigcup_{n=1}^{\infty} A_n \in \mathcal{F}$.

The pair $(\Omega, \mathcal{F})$ is called a *measurable space*. Let $(\Omega, \mathcal{F})$, $(E, \mathcal{E})$ be two measurable spaces. A mapping X from Ω into E such that for all $A \in \mathcal{E}$ the set $\{\omega : X(\omega) \in A\}$ belongs to $\mathcal{F}$ is called a *random variable*, or E-valued random variable or a measurable mapping.

Let $\mathcal{M}$ be a collection of subsets of Ω. The smallest σ-field on Ω, which contains $\mathcal{M}$, is denoted by $\sigma(\mathcal{M})$ and called the *σ-field generated by* $\mathcal{M}$. It is the intersection of all σ-fields on Ω containing $\mathcal{M}$. Analogously, let $(X_i)_{i \in I}$ be a family of mappings from Ω into $(E, \mathcal{E})$ then the smallest σ-field on Ω with respect to which all functions X_i are measurable is called the *σ-fields generated by* $(X_i)_{i \in I}$ and is denoted by $\sigma(X_i : i \in I)$. Given measurable spaces $(E_1, \mathcal{E}_1), (E_2, \mathcal{E}_2), \ldots, (E_k, \mathcal{E}_k)$, the σ-field $\mathcal{E}_1 \times \mathcal{E}_2 \times \ldots \times \mathcal{E}_k$ is the smallest σ-field of subsets of $E_1 \times E_2 \times \ldots \times E_k$ which contains all sets of the form $A_1 \times A_2 \times \ldots \times A_k$ where $A_i \in \mathcal{E}_i$, $i = 1, 2, \ldots, k$.

Let (E, ρ) be a metric space with metric ρ. The σ-field on E generated by closed subsets of E is called *Borel* and and is denoted by $\mathcal{B}(E)$. Metric spaces will be treated as measurable spaces with the σ-field of Borel sets. In particular random variables taking values in $\mathbb{R}^1$ will be called real random variables.

2.2. – Dynkin's π-λ theorem

Many arguments in the theory of stochastic processes depend on Dynkin's $\pi - \lambda$ theorem [21]. A collection $\mathcal{L}$ of subsets of Ω is said to be a π-system if it is closed under the formation of finite intersections: if $A, B \in \mathcal{L}$ then $A \cap B \in \mathcal{L}$. A collection $\mathcal{M}$ of subsets of Ω is a λ-system if it contains Ω, is closed under the formation of complements and of finite and countable disjoint unions:

1) $\Omega \in \mathcal{M}$.
2) If $A \in \mathcal{M}$ then $A^c \in \mathcal{M}$.
3) If $A_i \in \mathcal{M}$ and $A_i \cap A_j = \emptyset$ for $i \neq j$, $i, j = 1, 2, \ldots$ then $\bigcup_{i=1}^{\infty} A_i \in \mathcal{M}$.

THEOREM 2.2.1. *If a λ-system $\mathcal{M}$ contains a π-system $\mathcal{L}$, then $\mathcal{M} \supset \sigma(\mathcal{L})$.*

PROOF. Denote by $\mathcal{K}$ the smallest λ-system containing $\mathcal{L}$, equal to the intersection of all λ-systems containing $\mathcal{L}$. Then $\mathcal{K} \subset \sigma(\mathcal{L})$. To prove the opposite inclusion we show first that $\mathcal{K}$ is a π-system. Let $A \in \mathcal{K}$ and define $\mathcal{K}_A = \{B : B \in \mathcal{K} \text{ and } A \cap B \in \mathcal{K}\}$. It is easy to check that $\mathcal{K}_A$ is closed under the formation of complements and countable disjoint unions and if $A \in \mathcal{L}$ then $\mathcal{K}_A \supset \mathcal{L}$. Thus for $A \in \mathcal{L}$, $\mathcal{K}_A = \mathcal{K}$ and we have shown that if $A \in \mathcal{L}$ and $B \in \mathcal{K}$ then $A \cap B \in \mathcal{K}$. But this implies that $\mathcal{K}_B \supset \mathcal{L}$ and, consequently, $\mathcal{K}_B = \mathcal{K}$ for any $B \in \mathcal{K}$. It is now an easy exercise to show that if a π-system is closed under the formation of complements and countable disjoint unions then it is a σ-field. This completes the proof.   □

If E is a metric space then the family of all open (closed) subsets of E is a π-system. If $E = \mathbb{R}^d$ then the sets $\{y \in \mathbb{R}^d : y \leq x\}$, $x \in \mathbb{R}^d$, form a π-system.

If $(E, \mathcal{E})$ is a measurable space and μ a nonnegative function on $\mathcal{E}$ such that

1) $\mu(A) \in [0, \infty]$ for all $A \in \mathcal{E}$,
2) $\mu(\emptyset) = 0$,
3) If $A_n \in \mathcal{F}$, $A_n \cap A_m = \emptyset$ for $n \neq m$, then

$$\mu\left(\bigcup_n A_n\right) = \sum_{n=1}^{\infty} \mu(A_n).$$

Then μ is called a nonnegative measure or shortly a measure. If $\mu(E) = 1$ then the measure is called a *probability measure* or shortly *probability*.

The following Proposition will be frequently used.

PROPOSITION 2.2.2. *If two probability measures μ, ν on $(E, \mathcal{E})$ are identical on a π-system $\mathcal{L}$ then they are identical on $\sigma(\mathcal{L})$.*

PROOF. Let $\mathcal{M}$ be the collection of all $A \in \mathcal{E}$ such that

$$\mu(A) = \nu(A).$$

It is clear that $\mathcal{M}$ is a λ-system and therefore $\mathcal{M} \supset \sigma(\mathcal{L})$. $\qquad\square$

2.3. – Independence

We recall also the notion of independence playing an essential role in the theory of stochastic processes. Let $(\Omega, \mathcal{F}, \mathbb{P})$ be a probability space and let $(\mathcal{F}_i)_{i \in I}$ be a family of sub-σ-fields. These σ-fields are said to be independent if, for every finite subset $J \subset I$ and every family $(A_i)_{i \in J}$ such that $A_i \in \mathcal{F}_i$, $i \in J$,

$$(2.3.1) \qquad \mathbb{P}\left(\bigcap_{i \in J} A_i\right) = \prod_{i \in J} \mathbb{P}(A_i).$$

Random variables $(X_i)_{i \in I}$ are independent if the σ-fields $(\sigma(X_i))_{i \in I}$ are independent.

PROPOSITION 2.3.1. *If $\mathcal{L}_i$ are π-systems on Ω and $\mathcal{F}_i = \sigma(\mathcal{L}_i)$, $i \in I$, then the σ-fields $(\mathcal{F}_i)_{i \in I}$ are independent if for every finite set $J \subset I$ and sets $A_i \in \mathcal{L}_i, i \in I$,*

$$\mathbb{P}\left(\bigcap_{i \in J} A_i\right) = \prod_{i \in J} \mathbb{P}(A_i).$$

PROOF. Assume, without loss of generality, that $I = J = \{1, \ldots, n\}$. Let us fix the sets $A_i \in \mathcal{L}_i$, $i = 2, \ldots, n$, and denote by $\mathcal{M}_1$ the family of all sets $A_1 \in \mathcal{F}_1$ for which

$$(2.3.2) \qquad \mathbb{P}\left(\bigcap_{k=1}^{n} A_k\right) = \prod_{k=1}^{n} \mathbb{P}(A_k).$$

The family contains $\mathcal{L}_1$ and is a λ-system, therefore $\mathcal{M}_1 = \mathcal{F}_1$. Analogously, fix $A_1 \in \mathcal{F}_1$ and $A_i \in \mathcal{L}_i$, $i = 3, \ldots, n$, and denote by $\mathcal{M}_2$ the family of $A_2 \in \mathcal{F}_2$ for which (2.3.2) holds. Then $\mathcal{M}_2 = \sigma(\mathcal{L}_2) = \mathcal{F}_2$. Easy induction shows that (2.3.2) holds for all $A_i \in \mathcal{F}_i$, $i = 1, 2, \ldots, n$. $\qquad\square$

COROLLARY 2.3.2. *Let $X_1, X_2, \ldots$ be a sequence of real-valued random variables. The random variables $(X_i)_{i \in \mathbb{N}}$ are independent iff for arbitrary $n = 1, 2, \ldots$ and arbitrary real numbers $x_1, \ldots, x_n$:*

$$(2.3.3) \qquad \mathbb{P}(X_1 \le x_1, \ldots, X_n \le x_n) = \prod_{i=1}^{n} \mathbb{P}(X_i \le x_i).$$

PROOF. The σ-fields $\sigma(X_i)$ are generated by the events $\{X_i \leq x_i\}$ forming π-systems. $\square$

REMARK 2.3.3. Note that according to the definition (X_i) are independent if and only if for arbitrary n and arbitrary Borel subsets $\Gamma_1, \ldots, \Gamma_n$ of $\mathbb{R}^1$:

$$\mathbb{P}(X_1 \in \Gamma_1, \ldots, X_n \in \Gamma_n) = \prod_{i=1}^{n} \mathbb{P}(X_i \in \Gamma_i).$$

So the condition (2.3.3) is a real simplification of (2.3.1).

2.4. – Expectations and conditioning

Since $\mathbb{P}$ is a measure on $(\Omega, \mathcal{F})$, the Lebesgue integral,

$$(2.4.1) \qquad\qquad \int_{\Omega} X(\omega)\, \mathbb{P}(d\omega),$$

of a measurable, non-negative function X is well defined. It is defined also for those functions X for which either negative or positive parts has a finite integral. If

$$\int_{\Omega} |X(\omega)|\, \mathbb{P}(d\omega) < \infty,$$

then X is called integrable. In probability theory, the Lebesgue integral (2.4.1) is called expected value, or mean value or expectation of X and denoted as

$$\mathbb{E}\, X.$$

The law μ of a random variable X taking values in a measurable space $(E, \mathcal{E})$ is a probability measure on $(E, \mathcal{E})$ such that

$$\mu(\Gamma) = \mathbb{P}(\{\omega\,;\, X(\omega) \in \Gamma\}), \quad \Gamma \in \mathcal{E}.$$

If μ is the law of a random variable X then for arbitrary Borel, non-negative, or bounded, function $\phi : E \to \mathbb{R}^1$:

$$(2.4.2) \qquad\qquad \mathbb{E}(\phi(X)) = \int_{E} \phi(x)\, \mu(dx).$$

This identity is true for indicator functions just by the definition of μ. Therefore it holds for all measurable functions ϕ taking only a finite number of values. But then it holds for all nonnegative ϕ because they are limits of increasing sequences of functions taking a finite number of values. Decomposing arbitrary measurable ϕ into its positive and negative parts one shows that (2.4.2) is true in general. Identity (2.4.2) will be frequently used in the future.

Assume that $\mathcal{G}$ is a σ-field contained in $\mathcal{F}$ and X is a non-negative or integrable random variable then the conditional expectation

$$\mathbb{E}(X|\mathcal{G})$$

is a $\mathcal{G}$-measurable random variable Z such that for each $A \in \mathcal{G}$,

$$\mathbb{E}(X\, 1_A) = \mathbb{E}(Z\, 1_A).$$

We take for granted its elementary properties.

2.5. – Gaussian measures

A Gaussian measure μ on $\mathbb{R}^1$ is a probability measure either concentrated at one point, $\mu = \delta_m$, or having a density

$$\frac{1}{\sqrt{2\pi q}} e^{\frac{1}{2q}(x-m)^2}, \qquad x \in \mathbb{R}^1,$$

where $m \in \mathbb{R}^1$, $q > 0$.

If E is a Banach space then a probability measure μ on $(E, \mathcal{B}(E))$ is said to be Gaussian if and only if the law of an arbitrary linear functional $h \in E^*$ considered as a random variable on $(E, \mathcal{B}(E), \mu)$ is a Gaussian measure on $(\mathbb{R}^1, \mathcal{B}(\mathbb{R}^1))$.

In the case when $E = \mathbb{R}^n$, the Gaussian measures are characterised uniquely by the mean vector $m = (m_1, \ldots, m_n)$ and the covariance matrix $Q = (q_{jk})$, and denoted by $N(m, Q)$. If $m = 0$, one writes N_Q. Thus

$$\int_{\mathbb{R}^n} x_j \, N(m, Q)(dx) = m_j, \quad \int_{\mathbb{R}^n} (x_j - m_j)(x_k - m_k) \, N(m, Q)(dx) = q_{jk},$$

$$j, k = 1, \ldots, n.$$

If Q is positive definite $N(m, Q)$ has a density $g_{m,Q}$:

$$g_{m,Q} = \frac{1}{\sqrt{(2\pi)^n \det Q}} e^{-\frac{1}{2}\langle Q^{-1}(x-m),(x-m)\rangle}, \qquad x \in \mathbb{R}^n.$$

A random vector X is Gaussian if its law $\mathcal{L}(X)$ is a Gaussian measure. If $X = (X_1, \ldots, X_n)$ is Gaussian with values in $\mathbb{R}^n$, then the random variables X_i are independent if and only if the matrix Q is diagonal.

2.6. – Convergence of measures

A sequence (μ_n) of probability measures on a separable metric space E, with metric ρ, is said to be *weakly convergent* to a measure μ if for arbitrary continuous and bounded $\phi : E \to \mathbb{R}^1$:

$$(2.6.1) \qquad \int_E \phi(x) \, \mu_n(dx) \to \int_E \phi(x) \, \mu(dx) \qquad \text{as } n \to \infty.$$

If (μ_n) converges weakly to μ then one writes $\mu_n \Rightarrow \mu$. The space $C_b(E)$ of all bounded and continuous functions of E equipped with the supremum norm will be denoted by $C_b(E)$.

PROPOSITION 2.6.1. *For a given sequence (μ_n) there exists at most one weak limit.*

PROOF. It is enough to show that if for two measures μ, ν:

$$(2.6.2) \qquad \int_E \phi(x)\,\mu(dx) = \int_E \phi(x)\,\nu(dx)$$

for all $\phi \in C_b(E)$ then $\mu = \nu$. Let A be a closed set and ϕ_k, $k = 1, 2, \ldots$, the following continuous and bounded functions:

$$\phi_k(x) = \frac{1}{1 + k\rho(x, A)}, \qquad x \in E,$$

where $\rho(x, A) = \inf\{\rho(x, y)\,;\ y \in A\}$. Then $\phi_k \downarrow \chi_A$ as $k \to \infty$. Since (2.6.2) holds for all ϕ_k therefore it holds for the limit. Consequently $\mu(A) = \nu(A)$ for all closed sets. By Dynkin's $\pi - \lambda$ theorem $\mu = \nu$. $\qquad\square$

EXAMPLE 2.6.2. A centred Gaussian measure $\mathcal{N}_\sigma$ on $\mathbb{R}^d$, $\sigma > 0$, with the covariance matrix σI has the density

$$n_\sigma(x) = \frac{1}{\sqrt{(2\pi\sigma)^d}}\, e^{-\frac{|x|^2}{2\sigma}}, \qquad x \in \mathbb{R}^d.$$

If $\sigma \to 0$, then $\mathcal{N}_\sigma \Rightarrow \delta_{\{0\}}$, where $\delta_{\{0\}}$ is the probability measure concentrated in 0.

The following result is classical attributed to Helly. It will be proved in Chapter 10 as a consequence of more general results.

PROPOSITION 2.6.3. *Assume that* (μ_n) *is a sequence of probability measures on* $\mathbb{R}^d$ *and that for arbitrary* $\epsilon > 0$ *there exists* $R > 0$ *such that*

$$\mu_n(B(0, R)) \geq 1 - \epsilon, \qquad n = 1, 2, \ldots.$$

Then (μ_n) *contains a subsequence* (μ_{n_k}) *which converges weakly to a probability measure* μ.

2.7. – Characteristic functions

The characteristic function of a measure μ on $\mathbb{R}^d$ is defined as follows:

$$\phi_\mu(\lambda) = \int_E e^{i\langle \lambda, x \rangle}\,\mu(dx), \qquad \lambda \in \mathbb{R}^d.$$

Characteristic functions are continuous functions, in general, with complex values. They are denoted also as $\hat{\mu}$.

EXAMPLE 2.7.1. The characteristic function of $\mathcal{N}_\sigma$ is as follows:

$$\phi_{\mathcal{N}_\sigma}(\lambda) = e^{-\frac{\sigma}{2}|\lambda|^2}, \qquad \lambda \in \mathbb{R}^d.$$

Assume that $\sigma = 1$ and $\phi(\lambda) = \frac{1}{\sqrt{2\pi}} \int_{-\infty}^{+\infty} e^{i\lambda x} e^{-\frac{x^2}{2}} dx$, $\lambda \in \mathbb{R}^1$. Then

$$\phi'(\lambda) = \frac{1}{\sqrt{2\pi}} i \int_{-\infty}^{+\infty} e^{i\lambda x} x e^{-\frac{x^2}{2}} dx = -\frac{i}{\sqrt{2\pi}} \int_{-\infty}^{+\infty} e^{i\lambda x} (e^{-\frac{x^2}{2}})' dx$$

$$= \frac{i^2}{\sqrt{2\pi}} \lambda \int_{-\infty}^{+\infty} e^{i\lambda x} e^{-\frac{x^2}{2}} dx = -\lambda \phi(\lambda).$$

Therefore $\phi(\lambda) = e^{-\frac{\lambda^2}{2}}$, $\lambda \in \mathbb{R}^1$.

More generally, the Gaussian distribution $\mathcal{N}(m, Q)$ on $\mathbb{R}^d$ with covariance matrix Q and mean vector m has a characteristic function

$$e^{i\langle m,\lambda\rangle - \frac{1}{2}\langle Q\lambda,\lambda\rangle}, \qquad \lambda \in \mathbb{R}^d.$$

PROOF. For the proof assume that $m = 0$ and Q is invertible. Introduce the new basis given by the eigenvectors of Q. In the new basis:

$$\langle Q\lambda, \lambda \rangle = \sum_{k=1}^{d} q_k \lambda_k^2, \qquad \det Q = q_1 \ldots q_d, \qquad \langle Q^{-1}x, x \rangle = \sum_{k=1}^{d} \frac{1}{q_k} x_k^2$$

and therefore

$$\frac{1}{\sqrt{(2\pi)^d \det Q}} \int_{\mathbb{R}^d} e^{i\lambda x} e^{-\frac{1}{2}\langle Q^{-1}x,x\rangle} dx$$

$$= \frac{1}{\sqrt{(2\pi)^d \prod_{k=1}^{d} q_k}} \int_{\mathbb{R}^d} e^{i\sum_{k=1}^{d} x_k \lambda_k} e^{-\frac{1}{2}\sum_{k=1}^{d} \frac{1}{q_k} x_k^2} dx_1 \ldots dx_k$$

$$= e^{-\frac{1}{2}\sum_{k=1}^{d} q_k \lambda_k^2} = e^{-\frac{1}{2}\langle Q\lambda,\lambda\rangle}. \qquad \square$$

PROPOSITION 2.7.2. *Different probability measures define different characteristic functions.*

PROOF. If μ and ν are two measures and $\gamma \in \mathbb{R}^d$ then

$$e^{-i\langle\gamma,\lambda\rangle} \phi_\mu(\lambda) = \int_{\mathbb{R}^d} e^{i\langle\lambda, x-\gamma\rangle} \mu(dx).$$

Integrating both sides with respect to ν we have

$$\int_{\mathbb{R}^d} e^{-i\langle\gamma,\lambda\rangle} \phi_\mu(\lambda)\nu(d\lambda) = \int_{\mathbb{R}^d} \phi_\nu(x-\gamma)\mu(dx), \qquad \gamma \in \mathbb{R}^d.$$

In particular if $\nu = \mathcal{N}_\sigma$ one gets:

$$\frac{1}{\sqrt{(2\pi\sigma)^d}} \int_{\mathbb{R}^d} e^{-i\langle\gamma,\lambda\rangle} \phi_\mu(\lambda) e^{-\frac{|\lambda|^2}{2\sigma}} d\lambda = \int_{\mathbb{R}^d} e^{-\frac{\sigma}{2}|x-\gamma|^2} \mu(dx), \qquad \gamma \in \mathbb{R}^d.$$

$$\frac{1}{(2\pi)^d} \int_{\mathbb{R}^d} e^{-i\langle\gamma,\lambda\rangle} \phi_\mu(\lambda) e^{-\frac{|\lambda|^2}{2\sigma}} d\lambda = \int_{\mathbb{R}^d} n_{1/\sigma}(x-\gamma)\mu(dx), \qquad \gamma \in \mathbb{R}^d.$$

The right-hand side is the density of the convolution $\mathcal{N}_{1/\sigma} * \mu$ which converges weakly to μ as $\sigma \to \infty$. Since the left-hand side depends only on μ the result follows. $\qquad \square$

THEOREM 2.7.3. *A sequence (μ_n) of probability measures on $\mathbb{R}^d$ converges weakly to a probability measure μ if and only if $\phi_{\mu_n} \to \phi_\mu$ point-wise.*

PROOF. It is clear that if $\mu_n \Rightarrow \mu$ then $\phi_{\mu_n} \to \phi_\mu$ point-wise. For the opposite implication we need the following lemma.

LEMMA 2.7.4. *Assume that in a neighbourhood of 0 a sequence (ϕ_{μ_n}) of characteristic functions of measures (μ_n) converges to a function ϕ continuous at 0. Then for arbitrary $\epsilon > 0$ there exists $R > 0$ such that $\mu_n(B(0,R)) \geq 1 - \epsilon$ for $n = 1, 2, \ldots$.*

PROOF. It is enough to show that the result is true if $d = 1$. If $d > 1$ one considers projections of the measures (μ_n) on each of the coordinate axes. The sequences of projections are all tight by the one dimensional result and therefore the sequence (μ_n) is tight as well.

Let μ be a probability measure on $\mathbb{R}^1$ and r a positive number. Then

$$\frac{1}{2r}\int_{-r}^{r}(1-\phi_\mu(\lambda))\,d\lambda = \frac{1}{2r}\int_{-r}^{r}\left[\int_{-\infty}^{+\infty}(1-e^{i\lambda x})\,\mu(dx)\right]d\lambda$$

$$= \int_{-\infty}^{+\infty}\frac{1}{2r}\left[\int_{-r}^{r}(1-\cos\lambda x)\,d\lambda\right]\mu(dx) \geq \int_{|x|\geq\frac{2}{r}}\left(1-\frac{\sin xr}{xr}\right)\mu(dx)$$

$$\geq \int_{|x|\geq\frac{2}{r}}\left(1-\frac{1}{r|x|}\right)\mu(dx) \geq \frac{1}{2}\mu\left(x\,;\,|x|\geq\frac{2}{r}\right).$$

Assume that ϕ_{μ_n} converges to ϕ on an interval larger than $[-r,r]$. Since

$$\frac{1}{2r}\int_{-r}^{r}(1-\phi_{\mu_n}(\lambda))\,d\lambda \geq \frac{1}{2}\mu_n\left(x\,;\,|x|\geq\frac{2}{r}\right)$$

and $\lim_{\lambda\to0}\phi(\lambda) = 1$, $\lim_n\phi_{\mu_n}(\lambda) = \phi(\lambda)$, $\lambda \in [-r,r]$, therefore for $\epsilon > 0$ there exist n_0, r_0 such that for $n \geq n_0$ and $0 < r < r_0$

$$\frac{1}{2r}\int_{-r}^{r}(1-\phi_{\mu_n}(\lambda))\,d\lambda < \frac{\epsilon}{2}$$

and consequently for $n \geq n_0$, $\mu_n(x\,;\,|x|\geq\frac{2}{r}) \leq \epsilon$. Since an arbitrary finite sequence of measures on $\mathbb{R}^d$ is tight therefore the result follows. $\square$

By Helly's theorem the sequence (μ_n) contains a weakly convergent subsequence. Since all convergent subsequences converge to measures with the same characteristic functions the sequence is weakly convergent.

PROPOSITION 2.7.5. *If μ is a finite measure on $\mathbb{R}^d$ then*

$$\widehat{\mu}(\lambda) = \int_{\mathbb{R}^d}e^{i\langle\lambda,x\rangle}\mu(dx), \qquad \lambda \in \mathbb{R}^d,$$

is a uniformly continuous function.

PROOF. One can assume that μ is a probability measure. If this is the case then

$$
\begin{aligned}
|\widehat{\mu}(\lambda) - \widehat{\mu}(\eta)| &= \left| \int_{\mathbb{R}^d} e^{i\langle \lambda, x \rangle} (1 - e^{i\langle \eta - \lambda, x \rangle}) \mu(dx) \right| \\
&\leq \int_{\mathbb{R}^d} |1 - e^{i\langle \eta - \lambda, x \rangle}| \mu(dx) \\
&\leq \int_{|x| \leq R} |1 - e^{i\langle \eta - \lambda, x \rangle}| \mu(dx) + 2\mu\{x : |x| > R\}
\end{aligned}
$$

For $\epsilon > 0$ there exists $R > 0$ such that $2\mu\{x : |x| > R\} < \frac{\epsilon}{2}$. There exists also $\delta > 0$ such that if $|\eta - \lambda| < \delta$ then for all x, $|x| \leq R$, $|1 - e^{i\langle \eta - \lambda, x \rangle}| \leq \frac{\epsilon}{2\mu\{x : |x| \leq R\}}$. Consequently, if $|\eta - \lambda| < \delta$ then

$$
|\widehat{\mu}(\lambda) - \widehat{\mu}(\eta)| \leq \frac{\epsilon}{2} + \frac{\epsilon}{2\mu\{x : |x| \leq R\}} 2\mu\{x : |x| \leq R\} < \epsilon. \qquad \square
$$

PROPOSITION 2.7.6. *If μ_n, μ are probability measures and $\mu_n \Rightarrow \mu$ then $\widehat{\mu}_n \to \widehat{\mu}$ uniformly on bounded sets.*

PROOF. It is enough to show that if $\lambda_n \to \lambda$ then $\widehat{\mu}_n(\lambda_n) \to \widehat{\mu}(\lambda)$. Note that

$$
\begin{aligned}
|\widehat{\mu}_n(\lambda_n) - \widehat{\mu}(\lambda)| &\leq |\widehat{\mu}_n(\lambda_n) - \widehat{\mu}_n(\lambda)| + |\widehat{\mu}_n(\lambda) - \widehat{\mu}(\lambda)| \\
&\leq \int_{\mathbb{R}^d} |e^{i\langle \lambda_n, x \rangle} - e^{i\langle \lambda, x \rangle}| \mu_n(dx) + |\widehat{\mu}_n(\lambda) - \widehat{\mu}(\lambda)| \\
&\leq 2 \int_{|x| > R} \mu_n(dx) + \sup_{|x| \leq R} |1 - e^{i\langle \lambda_n - \lambda, x \rangle}| + |\widehat{\mu}_n(\lambda) - \widehat{\mu}(\lambda)| \\
&\leq I_1 + I_2 + I_3.
\end{aligned}
$$

The term I_3 tends to zero if $n \to +\infty$. Since the family (μ_n) is tight the first term I_1 can be made small by taking R sufficiently large, uniformly in n. Since $\lambda_n \to \lambda$, also the second term tends to zero if $n \to +\infty$. $\qquad \square$

CHAPTER 3

Kolmogorov's existence theorem

We prove here the Kolmogorov existence theorem and apply it to establish existence of Gaussian, Markov and Lévy processes.

3.1. – Carathéodory's and Ulam's theorems

A collection $\mathcal{E}_0$ of subsets of E is said to be a *field* or *algebra* if

1) $E \in \mathcal{E}_0$.
2) If $A \in \mathcal{E}_0$, then $A^c \in \mathcal{E}_0$.
3) If $A_1, \ldots, A_n \in \mathcal{E}_0$, then $\bigcup_{k=1}^{n} A_k \in \mathcal{E}_0$.

We take for granted the following result due to Carathéodory [8].

THEOREM 3.1.1 (Carathéodory). *If μ is a non-negative function on $\mathcal{E}_0$ such that*

1) $\mu(E) < \infty$. $\mu(\bigcup_{i=1}^{n} A_i) = \sum_{i=1}^{n} \mu(A_i)$ *if* $A_i \cap A_j = \emptyset$, $i \neq j$ *and* $A_i \in \mathcal{E}_0$.
2) *If* $A_i \in \mathcal{E}_0$, $A_i \cap A_j = \emptyset$ *for* $i \neq j$ *and* $\bigcup_i A_i \in \mathcal{E}_0$ *then*

$$\mu \left(\bigcup_{i=1}^{\infty} A_i \right) = \sum_{i=1}^{\infty} \mu(A_i).$$

Then there exists a unique extension of μ to a finite measure on the σ-field generated by $\mathcal{E}_0$.

REMARK 3.1.2. If Condition 1) is satisfied μ is called an *additive set function*. Condition 2) is equivalent to the following *continuity condition*,

2') If $A_i \in \mathcal{E}_0$ and $A_i \supset A_{i+1}$ for $i = 1, 2, \ldots$ and $\bigcap_{i=1}^{\infty} A_i = \emptyset$ then $\mu(A_i) \downarrow 0$ as $i \uparrow \infty$.

PROOF OF THE REMARK.

2) $\Rightarrow$ 2'). Since $\bigcap_{i=1}^{\infty} A_i = \emptyset$, $A_1 = \bigcup_{i=1}^{\infty}(A_i \setminus A_{i+1})$ and $A_i \setminus A_{i+1}$, $i = 1, 2 \ldots$, are disjoint elements of $\mathcal{E}_0$. Therefore

$$\mu(A_1) = \sum_{i=1}^{\infty}[\mu(A_i) - \mu(A_{i+1})] = \lim_n(\mu(A_1) - \mu(A_n))$$

and $\lim_n \mu(A_n) = 0$.

2') $\Rightarrow$ 2). For all $i = 1, 2, \ldots$, the sets $A_i' = \bigcup_{n=i}^{\infty} A_n$ belong to $\mathcal{E}_0$ and $\bigcap_i A_i' = \emptyset$. Therefore

$$\mu\left(\bigcap_{n=1}^{\infty} A_n\right) = \mu(A_1) + \ldots + \mu(A_i) + \mu\left(\bigcap_{n=i+1}^{\infty} A_n\right) = \mu(A_1) + \ldots + \mu(A_i) + \mu(A_{i+1}').$$

Since $\mu(A_{i+1}') \to 0$ as $i \to \infty$, the result follows. $\square$

The following result is due to Ulam [59].

THEOREM 3.1.3 (Ulam). *Assume that μ is a probability measure on a separable, complete, metric space. Then for arbitrary $\epsilon > 0$ there exists a compact set $K \subset E$ such that*

$$\mu(K) \geq 1 - \epsilon.$$

PROOF. Let (a_i) be a dense, countable sequence of elements of E. Then, for arbitrary $k = 1, 2, \ldots$,

$$\bigcup_{i=1}^{\infty} B\left(a_i, \frac{1}{k}\right) = E,$$

where $B(a, r) = \{y \in E : \rho(a, y) < r\}$, $a \in E, r > 0$. Therefore, for arbitrary k, there exists $n_k \in \mathbb{N}$ such that

$$\mu\left(\bigcup_{i=1}^{n_k} B\left(a_i, \frac{1}{k}\right)\right) > 1 - \frac{\epsilon}{2^k}.$$

The set $L = \bigcap_k \bigcup_{i=1}^{n_k} B(a_i, \frac{1}{k})$ is totally bounded and its closure K is compact. However

$$\mu(K) \geq 1 - \mu\left(\bigcup_{k=1}^{\infty}\left(\bigcup_{i=1}^{n_k} B\left(a_i, \frac{1}{k}\right)\right)^c\right)$$

$$\geq 1 - \sum_{k=1}^{\infty} \mu\left(\bigcup_{i=1}^{n_k} B\left(a_i, \frac{1}{k}\right)\right)^c > 1 - \sum_{k=1}^{\infty} \frac{\epsilon}{2^k} = 1 - \epsilon. \qquad \square$$

We have used the following elementary result left as an exercise.

PROPOSITION 3.1.4. *A metric space E is compact iff it is totally bounded*[1] *and complete.*

---

[1] A space E is totally bounded if for arbitrary $\epsilon > 0$ there exists a finite set $\{a_1, \ldots, a_n\}$ such that $\bigcup_{i=1}^{N} B(a_i, \epsilon) = E$.

3.2. – Kolmogorov's theorem

Let us recall that if $(\Omega, \mathcal{F})$ is a measurable space and $\mathbb{P}$ is a probability mesure on $\mathcal{F}$ then the collection $(\Omega, \mathcal{F}, \mathbb{P})$ is called a *probability space* and any family $X(t)$, $t \in \mathcal{T}$, of E-valued random variables defined on $(\Omega, \mathcal{F})$ is called a *stochastic process* on $\mathcal{T}$ or a *random family* on $\mathcal{T}$. Often $\mathcal{T}$ is a subset of $\mathbb{R}^1$. Traditionally if $\mathcal{T}$ is a subset of $\mathbb{R}^d$ with $d \geq 2$ then a stochastic process is called a *random field*.

As we have already said, a stochastic process X is usually described in terms of measures it induces on products of E, see (1.0.1). Namely if X is an E-valued process on $\mathcal{T}$ then for each sequence $(t_1, t_2, \ldots, t_k)$ of distinct elements of $\mathcal{T}$, $(X(t_1), \ldots, X(t_k))$ is a random variable with values in the Cartesian product $E \times E \times \ldots \times E$ equipped with the product σ-field $\mathcal{E} \times \mathcal{E} \times \ldots \times \mathcal{E}$. The probability measures on E^k:

$$\mu_{t_1, \ldots, t_k} = \mathcal{L}(X(t_1), \ldots, X(t_k)),$$

are called *finite-dimensional distributions* of X.

Note that the finite-dimensional distributions of a stochastic process $(X(t))_{t \in \mathcal{T}}$ satisfy two *consistency conditions*:

1) For any $A_i \in \mathcal{E}$, $i = 1, 2, \ldots, k$ and any permutation π of $(1, 2, \ldots, k)$:

$$\mu_{t_1, \ldots, t_k}(A_1 \times \ldots \times A_k) = \mu_{t_{\pi 1}, \ldots, t_{\pi k}}(A_{\pi 1} \times \ldots \times A_{\pi k}).$$

2) For any $A_i \in \mathcal{E}$, $i = 1, 2, \ldots, k - 1$,

$$\mu_{t_1, \ldots, t_{k-1}}(A_1 \times \ldots \times A_{k-1}) = \mu_{t_1, \ldots, t_k}(A_1 \times \ldots \times A_{k-1} \times E).$$

The following theorem is due to Kolmogorov.

THEOREM 3.2.1. *Assume that E is a separable, complete, metric space and $\mu_{t_1, \ldots, t_k}$ a family of distributions on E^k, $k \in \mathbb{N}$, satisfying 1) and 2), then there exists on some probability space $(\Omega, \mathcal{F}, \mathbb{P})$ a stochastic process $(X(t))_{t \in \mathcal{T}}$ having the $\mu_{t_1, \ldots, t_k}$ as its finite-dimensional distributions.*

The proof will be based on Carathéodory and Ulam theorems. We prove first a version of the Kolmogorov theorem of independent interest.

PROPOSITION 3.2.2. *Let $E_1, E_2, \ldots$ be a sequence of separable, complete, metric spaces and $\mu_1, \mu_2, \ldots$ probability measures on $\mathcal{B}(E_1)$, $\mathcal{B}(E_1) \times \mathcal{B}(E_2)$, $\ldots$. Then there exists a unique probability measure on $(E, \mathcal{B}(E))$ with $E = E_1 \times E_2 \times \ldots$ such that*

$$\mu(a_n \times E_{n+1} \times \ldots) = \mu_n(a_n), \qquad a_n \in \mathcal{B}(E_1) \times \ldots \mathcal{B}(E_n),$$

provided the following compatibility condition holds:

$$\mu_{n+1}(a_n \times E_{n+1}) = \mu_n(a_n), \qquad a_n \in \mathcal{B}(E_1) \times \ldots \mathcal{B}(E_n).$$

PROOF. Let $\mathcal{F}_0$ be the family of all sets of the form $a_n \times E_{n+1} \times \ldots$, $a_n \in \mathcal{B}(E_1) \times \ldots \mathcal{B}(E_n)$. One checks that $\mathcal{F}_0$ is a field and the set function

$$\mu(a_n \times E_{n+1} \times \ldots) = \mu_n(a_n), \qquad a_n \in \mathcal{B}(E_1) \times \ldots \mathcal{B}(E_n), \ n = 1, 2, \ldots,$$

satisfies condition 1) of the Carathéodory theorem. We check that also 2') holds. Let (A_n) be a decreasing sequence of elements of $\mathcal{F}_0$ such that $\bigcap_{n=1}^\infty A_n = \emptyset$ and assume that there exists $\epsilon > 0$ such that $\mu(A_n) \geq \epsilon$, $n = 1, 2, \ldots$. Without any loss of generality we can assume that

$$A_n = a_n \times E_{n+1} \times \ldots, \qquad \text{where } a_n \in \mathcal{B}(E_1) \times \ldots \mathcal{B}(E_n).$$

By Ulam's theorem there exists a compact set $b_n \subset a_n$ such that $\mu_n(a_n \setminus b_n) < \epsilon/(2^{n+1})$, $n = 1, 2, \ldots$. Denote $B_n = b_n \times E_{n+1} \times \ldots$ and define $C_n = \bigcap_{k=1}^n B_k$, $n = 1, 2, \ldots$. The sequence C_n is decreasing. Since

$$A_n \setminus C_n = A_1 \cap A_2 \cap \ldots \cap A_n \setminus B_1 \cap B_2 \cap \ldots \cap B_n \subseteq \bigcup_{k=1}^n (A_k \setminus B_k),$$

therefore

$$\mu(A_n \setminus C_n) \leq \sum_{k=1}^n \mu(A_k \setminus B_k) < \frac{\epsilon}{2}$$

and consequently $\mu(C_n) > \frac{\epsilon}{2}$, $n = 1, 2, \ldots$. Consequently $C_n, n = 1, 2, \ldots$ are nonempty sets. However $C_n = c_n \times E_{n+1} \times \ldots$ where c_n is a compact subset of $E_1 \times \ldots E_n$. Let (x_n) be any sequence $x_n = (x_{n1}, x_{n2}, \ldots) \in C_n$. By diagonal procedure one can extract a subsequence $x_{n_k} = (x_{n_k1}, x_{n_k2}, \ldots) \in C_{n_k}$ such that for each m, $x_{n_k m} \to x_{\infty m}$ as $k \to \infty$. But then $x_\infty = (x_{\infty 1}, x_{\infty 2}, \ldots) \in \bigcap_k C_{n_k} = \bigcap_n C_n \subset \bigcap_n A_n$ because all sets C_n are closed in E. This contradicts the identity $\bigcap_n A_n = \emptyset$. $\square$

PROOF OF THE KOLMOGOROV THEOREM. We define $\Omega = E^T$, $X(t, \omega) = \omega(t)$, $t \in T$ and $\omega \in \Omega$, $\mathcal{F} = \sigma(X(t), t \in T)$. Note that

$$\mathcal{F} = \bigcup_{\tau \subset T} \sigma(X(t), t \in \tau)$$

where the union is taken with respect to all countable subsets τ of T. Let

$$\mathcal{F}_0 = \bigcup_{\eta \subset T} \sigma(X(t), t \in \eta)$$

where the union is taken with respect to all finite subsets η of T. If $A \in \mathcal{F}_0$ then for some $t_1, \ldots, t_k \in T$ and $a \in \mathcal{B}(E^k)$, $A = \{\omega : (X(t_1, \omega), \ldots, X(t_k, \omega)) \in a\}$. We define a set function μ on $\mathcal{F}_0$ by the formula

$$\mu(A) = \mu_{t_1, \ldots, t_k}(a).$$

It is not difficult to see that Condition 1) of the Carathéodory theorem is satisfied. To check 2′) assume that $A_1 \supset A_2 \supset \ldots$ is a decreasing sequence of $\mathcal{F}_0$ such that $\bigcap_n A_n = \emptyset$. Without any loss of generality one can assume that $A_n \in \sigma(X(t), t \in \tau)$ where τ is a countable subset of T. Numbering elements of τ by natural numbers and taking into account that A_n are cylindrical sets we can assume that

$$A_n = a_n \times E^{\mathbb{N}}, \qquad \text{where } a_n \subset E^{m_n}, n = 1, 2, \ldots.$$

We can therefore apply the same reasoning as in the proof of the proposition to get that $\mu(A_n) \to 0$.   □

3.3. – Some applications

3.3.1. – Gaussian processes

As we already know a Gaussian measure on $\mathbb{R}^d$ is any probability measure on $\mathbb{R}^d$ whose characteristic function $\widehat{\mu}$ is of the form

$$\widehat{\mu}(\lambda) = e^{i \langle \lambda, m \rangle - \frac{1}{2} \langle Q\lambda, \lambda \rangle}, \qquad \lambda \in \mathbb{R}^d,$$

where m is a vector form $\mathbb{R}^d$ and Q is a non-negative definite matrix. The vector m and the matrix Q are the mean vector and the covariance matrix of μ. A random vector $Z = (Z_1, \ldots, Z_d)$ is Gaussian if its law $\mathcal{L}(Z)$ is a Gaussian measure. The random variables $Z_1, \ldots, Z_d$ are independent if and only if the covariance matrix is diagonal.

A real valued process $X(t)$, $t \geq 0$, is called *Gaussian* if for arbitrary numbers $t_1 < t_2 < \ldots < t_d$, $d = 1, 2, \ldots$, from the interval $[0, +\infty)$, the distribution of the random vector $(X(t_1), \ldots, X(t_d))$ is a Gaussian measure. Let X be Gaussian and define

$$(3.3.1) \quad m(t) = \mathbb{E}\, X(t), \quad q(s, t) = \mathbb{E}\,((X(s) - m(s))(X(t) - m(t))), \quad t, s, \in [0, +\infty).$$

The function q is called the *covariance* of the process X. Note that for arbitrary real numbers $\lambda_1, \lambda_2, \ldots \lambda_d$:

$$\mathbb{E}\left(\left(\sum_{j=1}^{d}(X(t_j) - m(t_j))\lambda_j\right)^2\right) = \sum_{j,k=1}^{d} q(t_j, t_k)\lambda_j \lambda_k$$

and therefore

$$(3.3.2) \qquad \sum_{j,k=1}^{d} q(t_j, t_k)\lambda_j \lambda_k \geq 0.$$

A function q for which (3.3.2) holds for arbitrary numbers $t_1 < t_2 < \ldots < t_d$, $d = 1, 2, \ldots$, from $[0, +\infty)$ and for arbitrary real numbers $\lambda_1, \lambda_2, \ldots \lambda_d$, $d = 1, 2, \ldots$, is called positive definite. Thus covariances of Gaussian processes are positive definite functions. We have the following result:

THEOREM 3.3.1. *For arbitrary function $m(t)$, $t \geq 0$, and arbitrary positive definite function $q(t, s)$, $t, s \geq 0$, there exists a Gaussian process for which (3.3.1) holds.*

PROOF. Let $\mu_{t_1,\dots,t_d}$ be the Gaussian measure with $m = (m(t_1), \dots, m(t_d))$ and $Q = (q(t_j, t_k))_{j,k=1\dots d}$. Let π be the transformation from $\mathbb{R}^d$ to $\mathbb{R}^{d-1}$ given by the formula:

$$\pi(x_1, \dots, x_d) = (x_1, \dots, x_{d-1}), \qquad x = (x_1, \dots, x_d) \in \mathbb{R}^d.$$

The consistency condition of Kolmogorov's theorem means that the image $\mu^{\pi}_{t_1,\dots,t_d}$ of the measure $\mu_{t_1,\dots,t_d}$ by the transformation π is the Gaussian measure $\mu_{t_1,\dots,t_{d-1}}$. Let us recall that in general if ν is the image of the measure μ by a transformation F then for a bounded measurable function ϕ

$$\int \phi(F(x)) \, \mu(dx) = \int \phi(y) \, \nu(dx).$$

Thus

$$\begin{aligned}
\widehat{\mu}^{\pi}_{t_1,\dots,t_d}(\lambda) &= \int_{\mathbb{R}^{d-1}} e^{i\langle \lambda, y \rangle} \mu^{\pi}_{t_1,\dots,t_d}(dy) \\
&= \int_{\mathbb{R}^d} e^{i\langle \lambda, \pi x \rangle} \mu_{t_1,\dots,t_d}(dx) \\
&= \int_{\mathbb{R}^d} e^{i\langle \pi^*\lambda, x \rangle} \mu_{t_1,\dots,t_d}(dx) \\
&= e^{i\langle \pi^*\lambda, m \rangle} e^{-\frac{1}{2}\langle Q\pi^*\lambda, \pi^*\lambda \rangle}.
\end{aligned}$$

Note that $\pi^*(\lambda_1, \dots, \lambda_{d-1}) = (\lambda_1, \dots, \lambda_{d-1}, 0)$, and therefore

$$\widehat{\mu}^{\pi}_{t_1,\dots,t_d}(\lambda) = e^{i\sum_{j=1}^{d-1} m(t_j)\lambda_j - \frac{1}{2}\sum_{j,k=1}^{d-1} q(t_j,t_k)\lambda_j\lambda_k}.$$

Consequently $\widehat{\mu}^{\pi}_{t_1,\dots,t_d}$ is the Gaussian measure on $\mathbb{R}^{d-1}$ with the mean vector $(m(t_1), \dots, m(t_{d-1}))$ and the covariance matrix $(q(t_j, t_k))_{j,k=1\dots d-1}$, as required. $\qquad\square$

3.3.2. – Wiener processes

The following function q:

$$q(t, s) = \min(t, s) =: t \wedge s, \qquad t, s \geq 0,$$

is positive definite. To see this set

$$\phi_t(x) = 1_{[0,t]}(x), \qquad x \geq 0.$$

Then $\phi_t \in L^2(0, +\infty)$ for every $t \geq 0$ and

$$q(t, s) = \langle \phi_t, \phi_s \rangle_{L^2(0,+\infty)}, \qquad t, s \geq 0.$$

Consequently

$$\sum_{j,k=1}^{d} q(t_j, t_k)\lambda_j\lambda_k = \sum_{j,k=1}^{d} \langle \phi_{t_j}, \phi_{t_k} \rangle \lambda_j\lambda_k$$

$$= \left\| \sum_{j=1}^{d} \phi_{t_j}\lambda_j \right\|^2 \geq 0.$$

By the theorem from the previous subsection there exists a Gaussian process $X(t)$, $t \geq 0$, such that

$$\mathbb{E}\,X(t) = 0, \qquad \mathbb{E}\,(X(t)X(s)) = t \wedge s, \qquad t, s \geq 0.$$

A Gaussian process X with continuous trajectories, mean value 0 and covariance function $q(t, s) = t \wedge s$, $t, x, \geq 0$, is called a standard *Wiener process*. Any Wiener process has independent increments in the sense that for any sequence of non-negative numbers $t_0 < t_1 < \ldots < t_d$, $d = 1, 2, \ldots$, the random variables

$$X(t_1) - X(t_0), \ldots, X(t_d) - X(t_{d-1})$$

are independent. In fact, by the very definition, the random vector

$$Z = (X(t_1) - X(t_0), \ldots, X(t_d) - X(t_{d-1}))$$

is Gaussian with the mean vector zero. The elements of its covariance matrix, out of the diagonal, can be easily calculated. For $j < k$

$$\mathbb{E}\,((X(t_j) - X(t_{j-1})(X(t_k) - X(t_{k-1})))$$
$$= t_j \wedge t_k - t_j \wedge t_{k-1} - t_{j-1} \wedge t_k + t_{j-1} \wedge t_{k-1}$$
$$= t_j - t_j - t_{j-1} + t_{j-1} = 0.$$

Since they vanish the increments are independent.

The constructed process is thus a process with independent increments with a prescribed covariance. Note however that the continuity of its trajectories does not follow from the Kolmogorov theorem and requires additional considerations.

3.3.3. – Markov processes

Let $(E, \mathcal{E})$ be a complete, separable metric space equipped with the σ-field $\mathcal{E} = \mathcal{B}(E)$ of Borel subsets of E. Let $\mathcal{P}(E)$ denote the set of probability measures on $(E, \mathcal{E})$. A *transition function* is a family of transformations P^t, $t \geq 0$, from E into $\mathcal{P}(E)$ which satisfy the *Chapman-Kolmogorov* equation:

$$P^{t+s}(x, \Gamma) = \int_E P^t(x, dy)P^s(y, \Gamma), \qquad x \in E, \ \Gamma \in \mathcal{E}, \ t, s, \geq 0.$$

The value $P^t(x, \Gamma)$ can be interpreted as the probability that a stochastic dynamical system starting from x will be in the set Γ at moment t. In a natural way one defines by induction, for any sequence of non-negative numbers $t_1 < t_2 < \ldots < t_d$, $d = 1, 2, \ldots$, the *probabilities of visiting sets* $\Gamma_1, \ldots, \Gamma_d$ at moments $t_1, \ldots, t_d$ *starting from* x as the functions $P^{t_1, \ldots, t_d} : E \to \mathcal{P}(E)$:

$$P^{t_1, \ldots, t_d}(x, \Gamma_1, \ldots, \Gamma_d) = \int_{\Gamma_1} P^{t_1}(x, dx_1) P^{t_2 - t_1, \ldots, t_d - t_{d-1}}(x_1, \Gamma_2, \ldots, \Gamma_d).$$

The Chapman-Kolmogorov equation implies that $P^{t_1, \ldots, t_{d-1}}$ is equal to the projection of $P^{t_1, \ldots, t_d}$ on the first $d - 1$ coordinates. By Kolmogorov's theorem there exists an E-valued process X, defined on a probability space $(\Omega, \mathcal{F}, \mathbb{P})$, such that

1) $X(0, \omega) = x$, $\omega \in \Omega$;
2) $\mathbb{P}(\{\omega : X(t_j, \omega) \in \Gamma_j, \ j = 1, \ldots, d\}) = P^{t_1, \ldots, t_d}(x, \Gamma_1, \ldots, \Gamma_d)$.

A stochastic process X for which the above two properties 1) and 2) hold is called a *Markov process with transition function* P^t. For more information see [30], [21], [5], [25], [56].

3.3.4. – *Lévy processes*

Let $E = \mathbb{R}^d$. If X is a stochastic process with values in $\mathbb{R}^d$ and for any sequence of non-negative numbers $t_0 < t_1 < \ldots < t_d$, $d = 1, 2, \ldots$, the random variables

$$X(t_1) - X(t_0), \ldots, X(t_d) - X(t_{d-1})$$

are independent than one says that X has *independent increments*. If, for all $t > s \geq 0$, the distribution of $X(t) - X(s)$ is identical with the distribution of $X(t - s)$, then X has *time homogeneous* increments. A process with independent and time homogeneous increments is called a *Lévy process* [43], [55], [2]. Lévy processes are closely related to the so called infinitely divisible families of measures.

A family μ_t, $t \geq 0$ of probability measures on $\mathbb{R}^d$ is *infinitely divisible* if

1) $\mu_0 = \delta_{\{0\}}$,
2) $\mu_{t+s} = \mu_t * \mu_s$ for all $t, s \geq 0$,
3) $\mu_t(\{x : \|x\| < r) \to 1$ as $t \downarrow 0$, for arbitrary $r > 0$.

If μ_t, $t \geq 0$ is an infinitely divisible family then the formula

(3.3.3) $P^t(x, \Gamma) = \mu_t(\Gamma - x),$ $\Gamma \in \mathcal{B}(R^d)$, $t \geq 0$, $x \in \mathbb{R}^d$,

defines a transition function. Consequently there exists a Markov process X, with the transition function P^t. The finite dimensional distributions $\mu_{t_1, t_2, \ldots, t_n}$ are determined by the identities:

(3.3.4)
$$\int_{\mathbb{R}^{nd}} \psi(x_1, \ldots, x_n) \, \mu_{t_1, t_2, \ldots, t_n}(dx_1, \ldots, dx_n)$$
$$= \int_{\mathbb{R}^{nd}} \psi(x_1, x_1 + x_2, \ldots, x_1 + \ldots + x_n) \mu_{t_1}(dx_1) \mu_{t_2 - t_1}(dx_2) \ldots \mu_{t_n - t_{n-1}}(dx_n),$$

valid for all bounded Borel ψ and for all sequences $0 = t_0 \leq t_1 < t_2 < \ldots < t_n$.

PROPOSITION 3.3.2. *Let $X(t)$, $t \geq 0$, be an $\mathbb{R}^d$-valued process with finite dimensional distributions $\mu_{t_1,\ldots,t_n}$ satisfying (3.3.4). Then for arbitrary Borel sets $\Gamma_1, \ldots, \Gamma_n$,*

$$\mathbb{P}(X(t_1) \in \Gamma_1, X(t_2) - X(t_1) \in \Gamma_2, \ldots X(t_n) - X(t_{n-1}) \in \Gamma_n)$$

(3.3.5)
$$= \prod_{k=1}^{n} \mathbb{P}(X(t_k) - X(t_{k-1}) \in \Gamma_k).$$

Moreover $\mathbb{P}(X(t) - X(s) \in \Gamma) = \mu_{t-s}(\Gamma)$, $t \geq s \geq 0$, Γ, Borel.

PROOF. Let $\phi_1, \ldots, \phi_n$ be bounded continuous functions on $\mathbb{R}^d$. Define $\psi(x_1, \ldots, x_n) = \phi_1(x_1)\phi_2(x_2 - x_1) \ldots \phi_n(x_n - x_{n-1})$, $x_1, \ldots, x_n \in \mathbb{R}^d$. Then

$$\mathbb{E}(\psi(X(t_1), \ldots, X(t_n))) = \int_{\mathbb{R}^{nd}} \psi(x_1, \ldots, x_n) \, \mu_{t_1,\ldots,t_n}(dx_1, \ldots, dx_n)$$

$$= \int_{\mathbb{R}^{nd}} \psi(z_1, z_1 + z_2 \ldots, z_1 + \ldots + z_n) \, \mu_{t_1}(z_1) \ldots \mu_{t_n - t_{n-1}}(dz_n)$$

$$= \int_{\mathbb{R}^{nd}} \phi_1(z_1)\phi_2(z_2) \ldots, \phi_n(z_n) \, \mu_{t_1}(z_1) \ldots \mu_{t_n - t_{n-1}}(dz_n)$$

$$= \prod_{k=1}^{n} \int_{\mathbb{R}^d} \phi_k(z_k) \, \mu_{t_k - t_{k-1}}(dz_k) = \prod_{k=1}^{n} \mathbb{E}(\phi_k(X(t_k - t_{k-1})))$$

$$= \prod_{k=1}^{n} \mathbb{E}(\phi_k(X(t_k) - X(t_{k-1}))). \qquad \square$$

Lévy processes have been intensively studied, see [55] and [2]. The complete description of their distributions, in terms of their characteristic functions, will be given in the chapter on Lévy processes. However explicit formulae for their distributions are known only in few cases. The most important ones are the following:

1. *Compound Poisson process* on $\mathbb{R}^d$ when

$$\mu_t = e^{-t\nu(\mathbb{R}^d)} \sum_{k=0}^{\infty} \frac{t^k}{k!} \nu^{*k},$$

and ν is a finite measure,
2. *Gaussian, Lévy processes* on $\mathbb{R}^d$. Here $\mu_t = N(ta, tQ)$, $a \in R^d$ and Q is a symmetric, non-negative, $d \times d$ matrix,
3. *Symmetric Cauchy processes* on $\mathbb{R}^d$. Here

$$\mu_t(dx) = \frac{\Gamma((d+1)/2)}{\pi^{(d+1)/2}} \frac{t}{(|x|^2 + t^2)^{(d+1)/2}} dx,$$

4. *Stable processes of order* $1/2$ on $\mathbb{R}^1_+$. Here

$$\mu_t(dx) = \frac{t}{\sqrt{2\pi}} \frac{1}{\sqrt{x^3}} e^{-\frac{t^2}{2x}} dx.$$

3.3.5. – Poisson processes

A stochastic Lévy process is called a *Poisson process* if its trajectories are increasing functions with non-negative integer values and with jumps of size 1. It is a special case of a compound Poisson process with measure $\nu = \lambda \delta_{\{0\}}$ and it is parametrised by parameter $\lambda > 0$.

The value $\pi(t)$ of the Poisson process at moment t can represent, for instance, the number of customers at a queue at moment t, or the number of car accidents in a given country in the time interval $[0, t]$.

CHAPTER 4

Doob's regularisation theorem

Kolmogorov's theorem allows to establish existence of a stochastic process X with prescribed finite dimensional distributions but does not imply any properties of its trajectories like continuity or right continuity. The aim of this chapter is to prove a powerful regularisation result of Doob which, in many cases, allows to prove existence of a *càdlàg version* of X that is a version which has right continuous trajectories with finite left-limits. The theorem and its proof is also a respectable pretext to introduce *martingales*, see [20], a powerful tool of the theory of stochastic processes. As an application we establish existence of càdlàg version of Lévy and Feller processes.

4.1. – Martingales and supermartingales

Let T be an arbitrary subset of $\mathbb{R}_+$ and let $(\Omega, \mathcal{F}, \mathbb{P})$ be a probability space and $(\mathcal{F}_t)_{t \in T}$ an increasing family of sub-σ-fields of $\mathcal{F}$. A family X_t, $t \in T$, of finite real-valued random variables adapted to the family $(\mathcal{F}_t)_{t \in T}$, i.e. each X_t is $\mathcal{F}_t$-measurable, $t \in T$, is said to be a *martingale* [61], respectively a supermartingale, a submartingale, with respect to $(\mathcal{F}_t)_{t \in T}$ if

1) X_t, $t \in T$, are integrable random variables.
2) If $s \leq t$, then for every event $A \in \mathcal{F}_s$

$$\int_A X_t \, d\mathbb{P} = \int_A X_s \, d\mathbb{P},$$

and respectively

$$\int_A X_t \, d\mathbb{P} \leq \int_A X_s \, d\mathbb{P},$$

$$\int_A X_t \, d\mathbb{P} \geq \int_A X_s \, d\mathbb{P}.$$

In terms of the conditional expectation property 2) can be rephrased as follows:

2') If $s \leq t$, then

$$\mathbb{E}\,(X_t|\mathcal{F}_s) = X_s, \qquad \mathbb{P} - \text{a.s.},$$

or respectively

$$\mathbb{E}\,(X_t|\mathcal{F}_s) \leq X_s, \qquad \mathbb{P} - \text{a.s.},$$
$$\mathbb{E}\,(X_t|\mathcal{F}_s) \geq X_s, \qquad \mathbb{P} - \text{a.s.}$$

Every real constant, respectively decreasing, increasing, function defined on T is a martingale, respectively a supermartingales, a submartingale.

LEMMA 4.1.1. *If (X_t) and (Y_t) are supermartingales relative to $(\mathcal{F}_t)$ and α, β are positive numbers then the processes $(\alpha X_t + \beta Y_t)$ and $(X_t \wedge Y_t)$ are also supermartingales. If (X_t) is a martingale, then $(|X_t|)$ is a submartingale.*

PROOF. Let $s \leq t$ and $A \in \mathcal{F}_s$. Since $\alpha \int_A X_s \, d\mathbb{P} \geq \alpha \int_A X_t \, d\mathbb{P}$ and $\beta \int_A Y_s \, d\mathbb{P} \geq \beta \int_A Y_t \, d\mathbb{P}$, therefore $\int_A (\alpha X_s + \beta Y_s) \, d\mathbb{P} \geq \int_A (\alpha X_t + \beta Y_t) \, d\mathbb{P}$.
Obviously,

$$\int_A (X_s \wedge Y_s) \, d\mathbb{P} = \int_{A \cap \{X_s < Y_s\}} X_s \, d\mathbb{P} + \int_{A \cap \{X_s \geq Y_s\}} Y_s \, d\mathbb{P}$$

and

$$A \cap \{X_s < Y_s\} \in \mathcal{F}_s, \qquad A \cap \{X_s \geq Y_s\} \in \mathcal{F}_s.$$

By virtue of the definition of a supermartingale, we obtain

$$\int_{A \cap \{X_s < Y_s\}} (X_s - X_t) \, d\mathbb{P} \geq 0, \qquad \int_{A \cap \{X_s \geq Y_s\}} (Y_s - Y_t) \, d\mathbb{P} \geq 0$$

and, consequently,

$$\int_A (X_s \wedge Y_s) \, d\mathbb{P} \geq \int_{A \cap \{X_s < Y_s\}} X_t \wedge Y_t \, d\mathbb{P} + \int_{A \cap \{X_s \geq Y_s\}} X_t \wedge Y_t \, d\mathbb{P} = \int_A X_t \wedge Y_t \, d\mathbb{P}.$$

If (X_t) is a martingale, then $(X_t \vee 0)$ and $(-X_t \vee 0)$ are submartingales, therefore $|X_t| = (X_t \vee 0) + (-X_t \vee 0)$ is a submartingale, too. $\square$

Let $(\Omega, \mathcal{F})$ be a measurable space and $(\mathcal{F}_t)_{t \in T}$ an increasing family of σ-fields, $\mathcal{F}_t \subset \mathcal{F}$, $t \in T$. A function $S : \Omega \to T$ is said to be a *stopping time*, relative to $(\mathcal{F}_t)$, if for every $t \in T$ the set $\{\omega : S(\omega) \leq t\}$ belongs to $\mathcal{F}_t$, i.e. the condition $S \leq t$ is a condition involving only what has happened up to and including time t. Let S be a stopping time. By $\mathcal{F}_S$ we denote the collection of events $A \in \mathcal{F}$ such that $A \cap \{S \leq t\} \in \mathcal{F}_t$ for all $t \in T$. It is easy to verify that $\mathcal{F}_S$ is a σ-field, the so-called *σ-fields of events prior to S*.

PROPOSITION 4.1.2.

a) *If T is a finite or countable subset of $\mathbb{R}$ then $S : \Omega \to T$ is a stopping time if and only if $\{\omega : S(\omega) = t\} \in \mathcal{F}_t$ for $t \in T$.*

b) *If S_1 and S_2 are stopping times then $S_1 \wedge S_2$ and $S_1 \vee S_2$ are again stopping times.*

c) *Any stopping time S is $\mathcal{F}_S$-measurable.*

d) *If $S_1 \leq S_2$ then $\mathcal{F}_{S_1} \subset \mathcal{F}_{S_2}$.*

PROOF. The properties $a), b), c)$ follow directly from the definitions. To prove $d)$ assume that $A \in \mathcal{F}_{S_1}$; then $\{S_2 \leq t\} \cap A = \{S_2 \leq t\} \cap \{S_1 \leq t\} \cap A$ because $\{S_2 \leq t\} \cap \{S_1 \leq t\} = \{S_2 \leq t\}$. Since $\{S_1 \leq t\} \cap A \in \mathcal{F}_t$ and $\{S_2 \leq t\} \in \mathcal{F}_t$ therefore $\{S_2 \leq t\} \cap \{S_1 \leq t\} \cap A \in \mathcal{F}_t$. $\qquad\square$

EXAMPLE 4.1.3. Let (X_n) be a sequence of random variables adapted to $(\mathcal{F}_n)$ and let $\mathcal{F}_\infty = \mathcal{F}$. Then

$$S = \begin{cases} \text{the least } n \text{ such that } X_n \geq a, \\ +\infty \text{ if } X_n < a \text{ for all } n = 1, 2, \ldots \end{cases}$$

is a stopping time. To see this fix a natural number k; then $\{S = k\} = \{X_1 < a, \ldots, X_{k-1} < a, X_k \geq a\} \in \sigma(X_1, \ldots, X_k) \subset \mathcal{F}_k$.

The following Doob's optional sampling theorem is of fundamental importance in the whole theory of martingales.

THEOREM 4.1.4. *Let $(X_n)_{n=1,\ldots,k}$ be a supermartingale (a martingale) relative to $(\mathcal{F}_n)_{n=1,\ldots,k}$. Let $S_1, S_2, \ldots, S_m$ be an increasing sequence of $(\mathcal{F}_n)$-stopping times with values in the set $\{1, \ldots, k\}$. The sequence $(X_{S_i})_{i=1,\ldots,m}$ is then also a supermartingale (a martingale) with respect to the σ-fields $\mathcal{F}_{S_1}, \mathcal{F}_{S_2}, \ldots, \mathcal{F}_{S_m}$.*

PROOF. Let $A \in \mathcal{F}_{S_1}$. We shall prove that $\int_A (X_{S_1} - X_{S_2}) \, d\mathbb{P} \geq 0$ in the supermartingale case and $\int_A (X_{S_1} - X_{S_2}) \, d\mathbb{P} = 0$ in the martingale case. If, for every ω, $S_2(\omega) - S_1(\omega) \leq 1$ then

$$\int_A (X_{S_2} - X_{S_1}) \, d\mathbb{P} = \sum_{r=1}^{k} \int_{A \cap \{S_1 = r\} \cap \{S_2 > r\}} (X_{r+1} - X_r) \, d\mathbb{P},$$

but $A \cap \{S_1 = r\}$ and $\{S_2 > r\}$ belong to $\mathcal{F}_r$, therefore $A \cap \{S_1 = r\} \cap \{S_2 > r\} \in \mathcal{F}_r$. The definition of a supermartingale (a martingale) implies that the desired inequality (equality) follows.

Let $r = 0, 1, \ldots, k$ and define stopping times $R_r = S_2 \wedge (S_1 + r)$. Then $S_1 = R_0 \leq R_1 \leq \ldots \leq R_k = S_2$ and $R_{i+1} - R_i \leq 1$, $i = 0, \ldots, k-1$. By virtue of the first part of the proof

$$\int_A X_{S_1} \, d\mathbb{P} \geq \int_A X_{R_1} \, d\mathbb{P} \geq \ldots \geq \int_A X_{R_k} \, d\mathbb{P} = \int_A X_{S_2} \, d\mathbb{P},$$

with equalities in the martingale case. $\qquad\square$

As the first application of the Doob theorem we establish the following Doob's inequalities.

THEOREM 4.1.5. *Let* $(X_n)_{n=1,\ldots,k}$ *be a supermartingale and* c *a non-negative constant. Then we have*

1) $c \, \mathbb{P}(\sup_n X_n \geq c) \leq \mathbb{E} \, X_1 - \int_{\{\sup_n X_n < c\}} X_k \, d\mathbb{P}.$
2) $c \, \mathbb{P}(\sup_n X_n \geq c) \leq \mathbb{E} \, X_1 + \mathbb{E} \, X_k^-.$
3) $c \, \mathbb{P}(\inf_n X_n \leq -c) \leq - \int_{\{\inf_n X_n \leq -c\}} X_k \, d\mathbb{P}.$
4) $c \, \mathbb{P}(\inf_n X_n \leq -c) \leq \mathbb{E} \, X_k^-.$

PROOF. To prove the inequalities 1), 2), define $S(\omega) = \inf\{n : X_n \geq c\}$, or $S(\omega) = k$ if $\sup_n X_n(\omega) < c$. S is a stopping time and $S \geq 1$. Thus, by Doob's optional sampling theorem, we obtain

$$\mathbb{E} \, X_1 \geq \mathbb{E} \, X_S$$

$$= \int_{\{\sup_n X_n \geq c\}} X_S \, d\mathbb{P} + \int_{\{\sup_n X_n < c\}} X_S \, d\mathbb{P}$$

$$\geq c \, \mathbb{P} \left(\sup_n X_n \geq c \right) + \int_{\{\sup_n X_n < c\}} X_k \, d\mathbb{P}$$

or equivalently

$$c \, \mathbb{P} \left(\sup_n X_n \geq c \right) \leq \mathbb{E} \, X_1 - \int_{\{\sup_n X_n < c\}} X_k \, d\mathbb{P}.$$

This is exactly inequality 1).

Since $-X_k \leq X_k^-$, inequality 2) follows, too.

To establish the relations 3) and 4), we introduce an analogous stopping time S, $S = \inf\{n : X_n \leq -c\}$ or $S = k$ if $\inf_n X_n > -c$. Since $S \leq k$ therefore

$$\mathbb{E} \, X_k \leq \mathbb{E} \, X_S$$

$$= \int_{\{\inf_n X_n \leq -c\}} X_S \, d\mathbb{P} + \int_{\{\inf_n X_n > -c\}} X_S \, d\mathbb{P}$$

$$\leq -c \, \mathbb{P} \left(\inf_n X_n \leq -c \right) + \int_{\{\inf_n X_n > -c\}} X_k \, d\mathbb{P}$$

These inequalities imply 3) and 4). □

Let $x = (x_1, \ldots, x_n)$ be a sequence of real numbers and let $a < b$. Let R_1 be the first of the numbers $1, 2, \ldots, n$ such that $x_{R_1} \leq a$, or n if there exist no such numbers. Let R_k be, for every even (respectively odd) integer $k > 1$, the first of the numbers $1, 2, \ldots, n$ such that $R_k > R_{k-1}$ and $x_{R_k} \geq b$ (respectively $x_{R_k} \leq a$), and if no such number exists we set $R_k = n$. In this way a sequence $R_1, R_2, \ldots$ is defined. The number $U_n(x; a, b)$ of upcrossings by the sequence x of the interval (a, b) is defined as the greatest integer k such that one actually has $x_{R_{2k-1}} \leq a$ and $x_{R_{2k}} \geq b$. If no such integer exists we put $U_n(x; a, b) = 0$.

THEOREM 4.1.6. *Let* $(X_m)_{m=1,...,n}$ *be a supermartingale relative to* $(\mathcal{F}_m)_{m=1,...,n}$ *and let* $a < b$ *be two real numbers. Then the following inequality holds:*

$$\mathbb{E}\, U_n(X; a, b) \le \frac{1}{b - a}\, \mathbb{E}\,((a - X_n)^+).$$

PROOF. Let $R_1, \ldots, R_{2l}$, $2l > n$, be the stopping times used in the definition of upcrossings and let $\Sigma_1 = (X_{R_2} - X_{R_1}) + \ldots (X_{R_{2l}} - X_{R_{2l-1}})$. Then

$$\Sigma_1 \ge (b - a)U_n(X; a, b) + (X_n - a) \wedge 0.$$

Indeed, assume that $U_n(X; a, b) = k$; then $\Sigma_1 = (X_{R_2} - X_{R_1}) + \ldots (X_{R_{2k+2}} - X_{R_{2k+1}})$. If $R_{2k+1} = n$ then $X_{R_{2k+2}} - X_{R_{2k+1}} = 0$, and if $R_{2k+1} < n$ and $R_{2k+2} = n$ then $X_{R_{2k+2}} - X_{R_{2k+1}} = X_n - X_{R_{2k+1}} \ge X_n - a$. Thus in both cases $X_{R_{2k+2}} - X_{R_{2k+1}} \ge (X_n - a) \wedge 0$.

Doob's theorem implies that

$$0 \ge \mathbb{E}\,(\Sigma_1) \ge (b - a)\, \mathbb{E}\, U_n(X; a, b) + \mathbb{E}\,((X_n - a) \wedge 0)$$

and therefore

$$(b - a)\, \mathbb{E}\, U_n(X; a, b) \le -\mathbb{E}\,((X_n - a) \wedge 0) = \mathbb{E}\,((a - X_n)^+). \qquad \square$$

4.2. – Regularisation theorem

Let $X(t)$, $t \ge 0$, be an E-valued stochastic process defined on $(\Omega, \mathcal{F}, \mathbb{P})$. An E-valued process $Y(t)$, $t \ge 0$, is said to be a modification of X if

$$\mathbb{P}(X(t) = Y(t)) = 1 \quad \text{for all} \quad t \ge 0.$$

It is clear that a modification Y of X has the same finite dimensional distributions as X. If there exists a modification Y of X which has, $\mathbb{P}$-almost surely, continuous trajectories, then we say that X has a continuous modification. If there exists a modification Y of X whose trajectories are right-continuous and have finite left limits then we say that X has a càdlàg (continu à droite et pourvu de limites à gauche) modification Y.

The following Doob's regularisation theorem states a condition under which a supermartingale has a càdlàg modification. Although it holds for supermartingales it has been successfully applied to wide classes of stochastic processes implying existence of their càdlàg versions. We will assume that the filtration $(\mathcal{F}_t)$ satisfies the usual conditions, that is

1) $\mathcal{F}$ is $\mathbb{P}$-complete.
2) $\mathcal{F}_0$ contains all $\mathbb{P}$-null sets of $\mathcal{F}$.

3) $(\mathcal{F}_t)$ is right-continuous:

$$\mathcal{F}_t = \bigcap_{u>t} \mathcal{F}_u =: \mathcal{F}_{t+}, \qquad t \geq 0.$$

THEOREM 4.2.1. *Assume that $X(t)$, $t \geq 0$, is a supermartingale such that for each $t \geq 0$ and $c > 0$*

(4.2.1) $$\lim_{s \to t, \, s > t} \mathbb{P}(|X(t) - X(s)| > c) = 0.$$

Then X has a càdlàg version.

4.2.1. – An application to Lévy processes

Before proving the result we deduce a simple application on regularity of Lévy processes.

THEOREM 4.2.2. *An arbitrary Lévy process X has a càdlàg version.*

PROOF. We can assume that X is a real valued Lévy process.

Let $\mathcal{G}_t = \sigma(X(s) : s \leq t)$, $t \geq 0$, $\mathcal{F}$ the completion of $\mathcal{G}_\infty = \sigma(X(s) : s \geq 0)$, and N_0 all sets of $\mathcal{F}$ with $\mathbb{P}$-measure zero. We set

$$\mathcal{F}_t = \bigcap_{s>t} \mathcal{G}_s \vee N_0, \qquad t \geq 0.$$

Then the family $(\mathcal{F}_t)$ satisfies the usual conditions and for each $s \leq t$ the increment $X(t) - X(s)$ is independent of $\mathcal{F}_s$.

For each number $u \in \mathbb{R}$ define a complex valued process

$$Z(t) = \frac{e^{iuX(t)}}{\mathbb{E}\, e^{iuX(t)}}, \qquad t \geq 0.$$

Then Z is a stochastic process for which real and imaginary parts are martingales. In fact

$$\mathbb{E}\,(Z(t)|\mathcal{F}_s) = \frac{1}{\mathbb{E}\, e^{iuX(t)}} \mathbb{E}\,(e^{iu(X(t)-X(s))+iuX(s)}|\mathcal{F}_s)$$

$$= \frac{1}{\mathbb{E}\, e^{iuX(t)}} \mathbb{E}\,(e^{iu(X(t)-X(s))}|\mathcal{F}_s)e^{iuX(s)}.$$

Since the random variable $X(t) - X(s)$ is independent of $\mathcal{F}_s$ we have that

$$\mathbb{E}\,(Z(t)|\mathcal{F}_s) = \frac{1}{\mathbb{E}\, e^{iuX(t)}} \mathbb{E}\,(e^{iu(X(t)-X(s))})e^{iuX(s)}.$$

It follows from the definition of Lévy processes, that

$$\mathbb{E}\,(e^{iu(X(t)-X(s))}) = \mathbb{E}\,(e^{iuX(t-s)}) = e^{-(t-s)\psi(u)},$$

where ψ is the Lévy-Khinchin exponent, and the lemma follows.

The continuity condition (4.2.1) from the theorem is satisfied and therefore for arbitrary $u \in \mathbb{R}$ the process $e^{iuX(t)}$, $t \geq 0$, has a càdlàg modification. It is possible to show that if, for a sequence (a_n) of real numbers, the limit $\lim_n e^{iua_n}$ exists for the set of all rational numbers u then the sequence (a_n) is convergent.

Let $X^u(t)$, $t \geq 0$, be a modification of $X(t)$, $t \geq 0$, such that the process

$$e^{iuX^u(t)}, \qquad t \geq 0,$$

is càdlàg. For each u we denote by Ω_u the set of those $\omega \in \Omega$ for which the trajectory of X^u is càdlàg. Denote by $\mathbb{Q}$ and, respectively $\mathbb{Q}_+$, the set of all rational and, respectively, non-negative rational numbers. Then the set

$$\widetilde{\Omega} = \bigcap_{u \in \mathbb{Q}_+} \bigcap_{v \in \mathbb{Q}_+} \bigcap_{r \in \mathbb{Q}_+} (\Omega_u \cap \{\omega : X^u(r, \omega) = X^v(r, \omega)\})$$

is of full measure. If we define

$$Y(t, \omega) = \begin{cases} X^u(t, \omega), & \omega \in \widetilde{\Omega} \text{ and } u \in \mathbb{Q}_+, \\ 0, & \omega \in \widetilde{\Omega}^c, \end{cases}$$

we get the required modification. $\qquad \qquad \square$

4.2.2. – Proof of the regularisation theorem

In the proof of Theore 4.2.1 we basically follow L. C. G. Rogers and D. Williams [53], vol. 1. To cope with measurability questions it is convenient to consider, at the beginning, functions defined on $\mathbb{Q}_+$ rather than on $\mathbb{R}_+$. We say that a function $x : \mathbb{Q}_+ \to \mathbb{R}$ has a càdlàg regularisation if the limits

$$\lim_{s>t, s\in\mathbb{Q}_+} x(s), \qquad \lim_{s<t, s\in\mathbb{Q}_+} x(s)$$

exist for all non-negative, respectively positive t. The following result is a direct consequence of the definition.

LEMMA 4.2.3. *If x has a càdlàg regularisation then $y : \mathbb{R}_+ \to \mathbb{R}$ given by the formula*

$$y(t) = \lim_{s>t, s\in\mathbb{Q}_+} x(s), \qquad t \geq 0,$$

is a càdlàg function.

The next lemma is of basic importance. To formulate it we need to generalize the definition of the upcrossing function $U^N(x; a, b)$. Namely the number $U^N(x; a, b)$ of upcrossings by the function x of the interval (a, b) is the supremum of $U_n((x(t_1), \dots, x(t_n)); a, b)$ with respect to all $n = 1, 2, \dots$ and arbitrary choice $0 \leq t_1 < t_2 < \dots < t_n \leq N$, $t_j \in \mathbb{Q}$.

LEMMA 4.2.4. *A function* $x : \mathbb{Q}_+ \to \mathbb{R}$ *has a càdlàg regularisation if and only if for all* $N \in \mathbb{N}$ *and* $a, b \in \mathbb{Q}$, $a < b$,

i) $\sup\{x(s) : s \in [0, N] \cap \mathbb{Q}_+\} < +\infty$,
ii) $U^N(x; a, b) < +\infty$.

PROOF. We prove for instance that if i) and ii) hold then x has a càdlàg regularisation. The other implication is similar. Thus assume i) and ii) hold but nevertheless for some $t \geq 0$

$$\limsup_{s > t,\, s \in \mathbb{Q}_+} x(s) > \liminf_{s > t,\, s \in \mathbb{Q}_+} x(s).$$

There exists rational numbers a, b such that

$$\liminf_{s > t,\, s \in \mathbb{Q}_+} x(s) < a < b \limsup_{s > t,\, s \in \mathbb{Q}_+} x(s).$$

Consequently for some decreasing sequences of rational numbers, $s_n > t$, $\bar{s}_n > t$, $n = 1, 2, \ldots$, converging to t:

$$x(s_n) \leq a, \quad x(\bar{s}_n) \geq b, \qquad n = 1, 2, \ldots.$$

This implies that $U^N(x; a, b,) = +\infty$ for $N > t$. Thus

$$\limsup_{s > t,\, s \in \mathbb{Q}_+} x(s) = \liminf_{s > t,\, s \in \mathbb{Q}_+} x(s) = \lim_{s > t,\, s \in \mathbb{Q}_+} x(s).$$

The condition i) implies that the limit is finite.   □

To prove the theorem define

$$\Gamma = \{\omega \in \Omega : X(s, \omega),\ s \in \mathbb{Q}_+,\ \text{has a càdlàg regularisation}\}.$$

Then

$$\Gamma = \bigcup_{N \in \mathbb{N};\, a, b \in \mathbb{Q}_+;\, a < b} \{\omega \in \Omega : U^N(X(\cdot, \omega); a, b) < +\infty,\ \sup_{s \in [0, N] \cap \mathbb{Q}_+} |X(s, \omega)| < +\infty\}.$$

It is easy to see that Γ is a measurable set and we define

$$Y(t, \omega) = \begin{cases} \displaystyle\lim_{s > t,\, s \in \mathbb{Q}_+} X(s, \omega), & \omega \in \Gamma, \\ 0, & \omega \in \Gamma^c. \end{cases}$$

We know that $Y(\cdot, \omega)$ is a càdlàg function for $\omega \in \Gamma$. We show now that $\mathbb{P}(\Gamma) = 1$. Let us fix N and rational numbers $a < b$. For an arbitrary sequence of rational numbers $0 = s_1 < s_2 < \ldots < s_{n-1} < s_n = N$ the sequence

$X(s_1), \dots, X(s_n)$ is a supermartingale. Therefore, by Doob's inequalities, for an arbitrary number $c \geq 0$:

$$\mathbb{P}\left(\sup_{k=1,\dots,n} X(s_k) \geq c\right) \leq \frac{1}{c}\left(\mathbb{E}\,X(0) + \mathbb{E}\,X^-(N)\right),$$

$$\mathbb{P}\left(\inf_{k=1,\dots,n} X(s_k) \leq -c\right) \leq \frac{1}{c}\mathbb{E}\,X^-(N).$$

From this, by considering an increasing family of increasing sequences $(s_1,\dots,s_n)$, covering $\mathbb{Q}_+ \cap [0, N]$, we obtain that

$$\lim_{c\uparrow+\infty} \mathbb{P}\left(\sup_{s\in\mathbb{Q}_+, s\leq N} |X(s)| \geq c\right) \leq \lim_{c\uparrow+\infty} \frac{1}{c}\left(\mathbb{E}\,X(0) + 2\mathbb{E}\,X^-(N)\right) = 0.$$

In a similar manner, by Doob's upcrossing theorem,

$$\mathbb{E}\,U^N((X(s_1), \dots, X(s_n)); a, b) \leq \frac{1}{b-a}\mathbb{E}\,(a - X(N))^+.$$

Therefore

$$\mathbb{E}\,U^N(X; a, b) \leq \frac{1}{b-a}\mathbb{E}\,(a - X(N))^+ < +\infty$$

and in particular

$$\mathbb{P}(U^N(X; a, b) < +\infty) = 1.$$

This way we have shown that $\mathbb{P}(\Gamma) = 1$.

It is clear that, for each $t \geq 0$, $Y(t)$ is measurable with respect to

$$\bigcap_{s>t} \sigma(X(u) \; : \; u \leq s) \vee N_0.$$

It remains to show that for each $t \geq 0$:

(4.2.2) $$\mathbb{P}(Y(t) = X(t)) = 1.$$

Let us fix $t \geq 0$ and let $t_n > t$, $n = 1, 2, \dots$, be a sequence of rational numbers converging to t. Then there exists a subsequence t_{n_k}, $k = 1, 2, \dots$, such that

$$\mathbb{P}\left(|X(t) - X(t_{n_k})| > \frac{1}{k}\right) \leq \frac{1}{k^2}.$$

Consequently by the Borel-Cantelli lemma

$$\lim_k X(t_{n_k}) = X(t), \qquad \mathbb{P} - \text{a.s.}$$

By the construction

$$\mathbb{P}\left(Y(t) = \lim_k X(t_{n_k})\right) = 1,$$

so (4.2.2) holds. $\qquad\qquad\qquad\qquad\qquad\qquad\qquad\qquad\qquad\qquad\qquad\qquad\square$

4.2.3. – An application to Markov processes

We deduce from Theorem 4.2.1 a general regularity result for Markov processes which implies the regularity of Lévy processes as a very special case.

Let E be a metric, separable, locally compact space and E_∂ the one-point compactification of E. If E is compact then ∂ is an isolated point. Let P^t be a transition function on E. Let $\mathcal{B}_b(E)$ be the space of all bounded Borel functions on E and $C_0(E)$ the space of all continuous functions on E having limit zero at ∂. The formula

$$P^t \phi(x) = \int_E P^t(x, dy)\, \phi(y), \qquad t \geq 0,\ x \in E,\ \phi \in \mathcal{B}_b(E),$$

defines a semigroup of operators, denoted also by P^t, on $\mathcal{B}_b(E)$. The transition function (P^t) is called Feller if (P^t) is a strongly continuous semigroup on $C_0(E)$. We can extend (P^t) to $\mathcal{B}_b(E_\partial)$ by setting

$$P^t(\partial, \{\partial\}) = 1, \quad P^t(x, \Gamma) = P^t(x, \Gamma \cap E), \quad \text{for } x \in E \text{ and } \Gamma \in \mathcal{B}(E_\partial).$$

The extended family is again a transition function. We have the following result.

THEOREM 4.2.5. *For arbitrary $x \in E$ there exists a càdlàg Markov process on E starting from x, with the Feller transition function (P^t).*

REMARK 4.2.6. As we noticed earlier there exists always a Markov process on E starting from $x \in E$ and with transition function P^t. To have a càdlàg version we first prove its existence on the compact space E_∂ rather than on a locally compact space.

REMARK 4.2.7. A transition semigroup (P^t) is said to be *stochastically continuous* if for all $x \in E$ and $\delta > 0$

$$\lim_{t \to 0} P^t(x, B(x, \delta)) = 1.$$

If P^t transforms $C_b(E)$ into $C_b(E)$ and is stochastically continuous, then for all $f \in C_b(E)$ and $x \in E$

$$\lim_{t \to 0} P^t f(x) = f(x).$$

Thus the strong Feller property is a slightly stronger condition than stochastic continuity.

PROOF OF THE THEOREM. Let A be the infinitesimal generator of the semigroup (P^t), regarded on $C_0(E)$, and R_λ, $\lambda > 0$, its resolvent. Then the domain $D(A)$ of A is dense in $C_0(E)$ and therefore also the image of R_1 is dense in $C_0(E)$. We can easily see that there exists a sequence (g_n) of non-negative functions such that the functions $f_n = R_1 g_n$, $n = 1, 2, \ldots$, separate points in E.

We extend the functions f_n to E_∂ by setting $f_n(\partial) = 0$. Now the transformation $H : E_\partial \to \mathbb{R}^\infty$ given by

$$H(x) = (f_1(x), f_2(x), \dots)$$

is continuous onto a compact set and therefore with a continuous inverse (defined on the image). Consequently $x_n \to x$ in E_∂ if and only if $f_k(x_n) \to f_k(x)$ for each k, as $n \to +\infty$. Let now $X(t)$, $t \geq 0$, be a Markov process on E starting from x. Then for arbitrary $k = 1, 2, \dots$ the process $e^{-t} f_k(X(t))$, $t \geq 0$, is a supermartingale, first with respect to the natural filtration $\sigma(X(s) : s \leq t)$, $t \geq 0$, and then also with respect to the extended versions. This follows from the formula, valid for all $f \in \mathcal{B}_b(E)$:

$$\mathbb{E}\left(f(X(t)) \mid \sigma(X(u) : u \leq s)\right) = P^{t-s} f(X(s)), \qquad s < t,$$

which can be checked using the π-systems technique. By a similar argument as for Lévy processes we infer that there exists an E_∂-valued càdlàg modification Y of X. To establish existence of an E-valued càdlàg version we need the following lemma, left as an exercise.

LEMMA 4.2.8. *Assume that Z is a càdlàg, non-negative supermartingale and define*

$$S(\omega) = \inf\{t \geq 0 : Z(t, \omega) = 0 \text{ or } Z(t^-, \omega) = 0\}.$$

Then

$$\mathbb{P}(Z(s) = 0 \text{ for all } s \geq S) = 1.$$

By the lemma, for arbitrary $t \geq 0$, if either $Y(t^-) = \partial$ or $Y(t) = \partial$, then $Y(s) = \partial$ for $s \geq t$. Since, for $x \in E$, the transition probabilities $P^t(x, \cdot)$ are supported by E, we see that $\mathbb{P}(Y(t) \in E \text{ for all } t \geq 0) = 1$. $\qquad\square$

CHAPTER 5

Wiener's approach

The first rigorous proof of the existence of the Wiener process was given in 1923 by N. Wiener [62]. It was based on Daniell's method [13] of constructing measures on infinite dimensional spaces. In 1932, N. Wiener together with R. E. Paley [51], gave an explicit construction of the Wiener process using Fourier series expansions and assuming only existence of a sequence of independent, identically distributed Gaussian random variables. We will present the essence of the method by describing the so-called Lévy-Ciesielski construction of the Wiener process. We describe also an explicit construction of the Poisson process. At the end of the chapter direct construction of Steinhaus, of an arbitrary sequence of independent random variables, is given.

5.1. – Hilbert space expansions

We start from some elementary facts about Hilbert spaces. Let H be a separable Hilbert space with scalar product $\langle h, g \rangle$ and the norm:

$$\|h\| = \sqrt{\langle h, h \rangle}, \qquad h \in H.$$

Let $h_1, h_2, \ldots$ be an orthonormal complete basis in H. Thus

$$|h_j| = 1, \quad \langle h_j, h_k \rangle = 0, \qquad j = 1, 2, \ldots, \ j \neq k,$$

and for arbitrary $x \in H$,

$$x = \sum_{j=1}^{+\infty} \langle x, h_j \rangle h_j, \qquad x \in H.$$

The following *Parseval identity* easily follows:

$$\langle x, y \rangle = \sum_{j=1}^{+\infty} \langle x, h_j \rangle \langle y, h_j \rangle, \qquad x \in H.$$

Take now $H = L^2(\Omega, \mathcal{F}, \mathbb{P})$ and assume that for each $t \geq 0$, $X(t) \in H$. If random variables $\xi_1, \xi_2, \ldots$ form in H an orthonormal complete basis then we arrive at the *first expansion*:

$$X(t) = \sum_{j=1}^{+\infty} \xi_j \langle X(t), \xi_j \rangle_{L^2(\Omega)} = \sum_{j=1}^{+\infty} \xi_j e_j(t), \qquad t \geq 0,$$

with the convergence, for each $t \geq 0$, in the sense of the space H.

If one takes $H = L^2(0, T)$ and if $X(\cdot, \omega) \in H$ and $e_1, e_2, \ldots$ is an orthonormal complete basis in $L^2(0, T)$, then we get the *second expansion*:

$$X(t, \omega) = \sum_{j=1}^{+\infty} \langle X(\cdot, \omega), e_j \rangle_{L^2(0,T)} e_j(t) = \sum_{j=1}^{+\infty} \xi_j(\omega) e_j(t), \qquad t \in [0, T].$$

Here, for each $\omega \in \Omega$, the convergence is in the $L^2(0, T)$ sense. Assume, for instance, that

$$\mathbb{E}X(t) = 0, \quad q(t, s) = \mathbb{E}(X(t)X(s)), \qquad t, s \in [0, T].$$

Define

$$Qx(t) = \int_0^T q(t, s)x(s)\, ds, \qquad t \in [0, T], \ x \in H = L^2(0, T),$$

and let e_j be an orthonormal complete basis in H, such that

$$Qe_j = \lambda_j e_j, \quad j = 1, 2, \ldots, \qquad \lambda_j > 0, \quad j = 1, 2, \ldots.$$

Then

$$\xi_j = \int_0^T X(t)e_j(t)\, dt, \qquad j = 1, 2, \ldots,$$

and

$$\mathbb{E}(\xi_j \xi_k) = \int_0^T \int_0^T q(t, s)e_j(t)e_k(s)\, dt ds = \lambda_j \delta_{j-k}.$$

If, in particular, $X(t) = W(t)$, $t \in [0, 1]$, is a Wiener process, then

$$q(t, s) = \mathbb{E}(W(t)W(s)) = t \wedge s,$$

and

$$e_j(t) = \sqrt{2}\sin\left[\left(j + \frac{1}{2}\right)\pi t\right], \quad \lambda_j = \frac{1}{\pi^2\left(j + \frac{1}{2}\right)^2}, \qquad j = 0, 1, 2, \ldots.$$

Therefore, for each $\omega \in \Omega$,

$$(5.1.1) \qquad W(t, \omega) = \sqrt{2}\sum_{j=0}^{+\infty}\xi_j(\omega)\frac{\sin\left[\left(j + \frac{1}{2}\right)t\right]}{\left(j + \frac{1}{2}\right)\pi}, \qquad t \in [0, 1],$$

and $\xi_0, \xi_1, \xi_2, \ldots$ are independent Gaussian random variables such that $\mathbb{E}\xi_j = 0$, $\mathbb{E}\xi_j^2 = 1$. This expression was found by Wiener. The difficult thing however is to prove that the series (5.1.1) defines a process with continuous paths. This amounts to the proof that for almost all $\omega \in \Omega$ the series or its subseries, converges uniformly. Instead of proving this we will present a similar and an elegant construction due to P. Lévy and Z. Ciesielski, see [43] and [10].

5.2. – The Lévy-Ciesielski construction of Wiener's process

In the Lévy-Ciesielski construction, the essential role is played by a Haar system connected with a dyadic partition of the interval $[0, 1]$.

Namely let $h_0 \equiv 1$, and if $2^n \leq k < 2^{n+1}$ then

$$h_k(t) = \begin{cases} 2^{n/2} & \text{if}\dfrac{k - 2^n}{2^n} \leq t < \dfrac{k - 2^n}{2^n} + \dfrac{1}{2^{n+1}}, \\[2ex] -2^{n/2} & \text{if}\dfrac{k - 2^n}{2^n} + \dfrac{1}{2^{n+1}} \leq t < \dfrac{k - 2^n}{2^n} + \dfrac{1}{2^n}, \end{cases}$$

$$h_k(1) = 0.$$

The system h_k, $k = 0, 1, \ldots$ forms an orthonormal and complete basis in the space $L^2([0, 1])$.

THEOREM 5.2.1. *Let $(\xi_k)_{k=0,1,\ldots}$ be a sequence of independent random variables normally distributed with mean 0 and covariance 1, defined on a probability space $(\Omega, \mathcal{F}, \mathbb{P})$. Then, for $\mathbb{P}$-almost all $\omega \in \Omega$, the series*

$$\sum_{k=0}^{+\infty}\xi_k(\omega)\int_0^t h_k(s)\,ds = W(t, \omega), \qquad t \in [0, 1],$$

is uniformly convergent on $[0, 1]$ and defines a Wiener process on $[0, 1]$.

LEMMA 5.2.2. *Let $\epsilon \in (0, 1/2)$ and $M > 0$. If $|a_k| \leq M k^\epsilon$ for $k = 1, 2, \ldots$ then the series $\sum_{k=0}^{+\infty} a_k \int_0^t h_k(s)\,ds$ is uniformly convergent on the interval $[0, 1]$.*

PROOF. If $2^n \leq k < 2^{n+1}$ then the *Schauder functions* $S_k(t) = \int_0^t h_k(s) \, ds$, $t \in [0, 1]$, are non-negative, have disjoint supports and are bounded from above by $2^{-(n+1)} 2^{\frac{n}{2}} = 2^{-\frac{n}{2}-1}$. Let us denote $b_n = \max(|a_k| : 2^n \leq k < 2^{n+1})$; then

$$\sum_{2^n \leq k < 2^{n+1}} |a_k| \, S_k(t) \leq b_n 2^{-\frac{n}{2}-1}$$

for all $t \in [0, 1]$ and $n = 0, 1, \dots$. Thus the condition

$$\sum_{n=0}^{+\infty} b_n 2^{-\frac{n}{2}} < +\infty$$

is sufficient for the uniform convergence of the series

$$\sum_{n=0}^{+\infty} \sum_{2^n \leq k < 2^{n+1}} |a_k| \, S_k(t)$$

and therefore for the uniform convergence of

$$\sum_{k=0}^{+\infty} a_k \, S_k(t)$$

too. From the inequalities $|a_k| \leq M k^\epsilon$ it follows that $b_n \leq 2^\epsilon M 2^{n\epsilon}$ for all $n = 0, 1, \dots$ and, consequently,

$$\sum_{n=0}^{+\infty} b_n 2^{-\frac{n}{2}} \leq 2^\epsilon M \sum_{n=0}^{+\infty} 2^{n(-\frac{1}{2}+\epsilon)} < +\infty. \qquad \square$$

LEMMA 5.2.3. *Let $(\xi_k)_{k=0,1,\dots}$ be a sequence of normally distributed random variables with mean 0 and covariance 1. Then, with probability one, the sequence*

$$\left(\frac{|\xi_k|}{\sqrt{\log k}} \right)_{k=2,3,\dots}$$

is bounded.

PROOF. Let c be a fixed positive number, then

$$\mathbb{P}(|\xi_k| \geq c) = \frac{2}{\sqrt{2\pi}} \int_c^{+\infty} e^{-x^2/2} \, dx \leq \frac{2}{\sqrt{2\pi}} \int_c^{+\infty} \frac{x}{c} e^{-x^2/2} \, dx \leq \frac{2}{c\sqrt{2\pi}} e^{-c^2/2}.$$

From this we obtain that for $c > \sqrt{2}$

$$\sum_{k=2}^{+\infty} \mathbb{P}(|\xi_k| \geq c\sqrt{\log k}) \leq \frac{2}{\sqrt{2\pi}} \sum_{k=2}^{+\infty} \frac{k^{-c^2/2}}{c\sqrt{\log k}} < +\infty.$$

Therefore, if $c > \sqrt{2}$ then, with probability one, only for a finite number of k, we have $|\xi_k| \geq c\sqrt{\log k}$.   □

PROOF OF THE THEOREM. Lemma 5.2.2 and Lemma 5.2.3 imply that the series

$$W(t, \omega) := \sum_{k=0}^{+\infty} \xi_k(\omega) S_k(t)$$

is for almost all ω uniformly convergent on the interval $[0, 1]$. Since the functions S_k are continuous and $S_k(0) = 0$ for $k = 0, 1, \ldots$ the constructed process starts from 0 and has continuous paths. As the limit of Gaussian processes, W is also a Gaussian process. Moreover

$$\mathbb{E}(W(t)W(s)) = \sum_{k=0}^{+\infty} \int_0^t h_k(u)\, du \int_0^s h_k(u)\, du = \sum_{k=0}^{+\infty} \langle h_k, \phi_t \rangle_{L^2(0,1)} \langle h_k, \phi_s \rangle_{L^2(0,1)},$$

where $\phi_t = 1_{[0,t]}$. Therefore by Parseval's identity

$$\mathbb{E}(W(t)W(s)) = \langle \phi_t, \phi_s \rangle_{L^2(0,1)} = t \wedge s.$$ □

5.2.1. – Construction of a Poisson process

A non-negative random variable ξ has exponential distribution with parameter $\alpha > 0$ if

$$\mathbb{P}(\xi > t) = e^{-\alpha t}, \quad t \geq 0.$$

The distribution of ξ has a density $g(t) = \alpha e^{-\alpha t}$, $t \geq 0$.

We have the following result.

PROPOSITION 5.2.4. *Assume that (ξ_k) is a sequence of independent, exponentially distributed random variables with parameter α. If*

$$\Pi(t) = 0 \quad \text{if} \quad \xi_1 > t$$
$$= k \quad \text{if} \quad \xi_1 + \ldots + \xi_k \leq t < \xi_1 + \ldots + \xi_{k+1},$$

then Π is a Poisson process with parameter α.

PROOF. It is enough to show that

(*i*) $\mathbb{P}(\Pi(t) = k) = e^{-\alpha t} \frac{(\alpha t)^k}{k!}$.

(*ii*) For arbitrary $k = 1, 2, \ldots$ and $t_0 = 0 < t_1 < \ldots < t_k$:

$$\mathbb{P}(\Pi(t_i) - \Pi(t_{i-1}) = k_i, i = 1, \ldots, nk) = \prod_{i=1}^{k} \mathbb{P}(\Pi(t_i) - \Pi(t_{i-1}) = k_i)$$

$$= \prod_{i=1}^{k} \mathbb{P}(\Pi(t_i - t_{i-1}) = k_i).$$

Let us recall that if ξ and η are independent random variables in a normed space E, say $E = \mathbb{R}^1$ or $E = \mathbb{R}^d$, with the laws $\mu = \mathcal{L}(\xi)$, $\nu = \mathcal{L}(\eta)$ then their sum $\zeta = \xi + \eta$ has the law $\sigma = \mathcal{L}(\xi + \eta)$ equal to the convolution of μ and ν:

$$\sigma(\Gamma) = \int_E \mu(\Gamma - y)\,\nu(dy), \qquad \Gamma \in \mathcal{B}(E).$$

In fact if ξ and η are independent then

$$\mathbb{P}(\xi + \eta \in \Gamma) = \mathbb{E}(\chi_\Gamma(\xi + \eta)) = \int_E \chi_\Gamma(x + y)\mu(dx)\nu(dy)$$
$$= \int_E \left[\int_E \chi_\Gamma(x + y)\mu(dx)\right]\nu(dy).^{(1)}$$

But $\int_E \chi_\Gamma(x+y)\mu(dx) = \mu(\Gamma - y)$, $y \in E$, and the required identity holds.

If random variables $\xi_1, \ldots, \xi_k$ are independent, exponentially distributed with parameter α then, by an easy inductive argument, the distribution of $\xi_1 + \ldots + \xi_k$ is a measure on $\mathbb{R}^1$ with the following density g_k:

$$g_k(x) = \alpha \frac{(\alpha x)^{k-1}}{(k-1)!} e^{-\alpha x}, \qquad x > 0, k = 1, 2, \ldots .$$

Note that for $k = 1, 2, \ldots$

$$\mathbb{P}(\Pi(t) = k) = \mathbb{P}(X_1 + \ldots + X_k \le t < X_1 + \ldots + X_{k+1})$$
$$= \iint_{\substack{0<y\le t<y+z \\ 0<z}} g_k(y)(\alpha e^{-\alpha z})dydz = \frac{\alpha^{k+1}}{(k-1)!} \iint_{\substack{0<y\le t \\ t-y<z}} y^{k-1}e^{-\alpha y}e^{-\alpha z}dydz$$
$$= \frac{\alpha^k}{(k-1)!} \int_{0<y\le t} y^{k-1}e^{-\alpha y}e^{-\alpha(t-y)}dy = \frac{\alpha^k}{(k-1)!}e^{-\alpha t}\int_0^t y^{k-1}dy = \frac{(\alpha t)^k}{k!}e^{-\alpha t}.$$

This proves (i). The identity (ii) is equivalent to the independence of the random variables $\Pi(t_i) - \Pi(t_{i-1})$, $i = 1, \ldots, k$ and will follow from some elementary properties of exponentially distributed random variables which will be developed now.

Let $\xi_1, \xi_2, \ldots$ be a sequence of independent random variables such that $\mathbb{P}(\xi_m = 1) = p$, $\mathbb{P}(\xi_m = 0) = 1 - p$, $m = 1, 2\ldots$. Let $T = \min\{m\,;\, \xi_m = 1\}$. Then $\mathbb{P}(T = k) = (1 - p)^{k-1}p$, $k = 1, 2\ldots$. Note that the random variable T is the *waiting time for the first success* in the sequence $\xi_1, \xi_2, \ldots$.

PROPOSITION 5.2.5. *Denote by μ_n the distribution of the random variable T/n where $p = \alpha/n$, $n = 1, 2\ldots$ and α is a positive constant. Then (μ_n) converges weakly to the exponential distribution with parameter α.*

$^{(1)}$ We used the notation $\chi_\Gamma(z) = 1$ if $z \in \Gamma$, $\chi_\Gamma(z) = 0$ if $z \notin \Gamma$.

PROOF. The characteristic function of the distribution of T is

$$\mathbb{E}(e^{i\lambda T}) = \sum_{k=1}^{\infty} e^{i\lambda k}(1-p)^{k-1}p = pe^{i\lambda}\frac{1}{1-(1-p)e^{i\lambda}}, \qquad \lambda \in \mathbb{R}^1.$$

Note that

$$\widehat{\mu}_n(\lambda) = \frac{\alpha}{n}e^{i\frac{\lambda}{n}}\frac{1}{1-\left(1-\frac{\alpha}{n}\right)e^{i\frac{\lambda}{n}}}, \qquad \lambda \in \mathbb{R}^1$$

and $\widehat{\mu}_n(\lambda) \to \frac{\alpha}{\alpha-i\lambda}$, $\lambda \in \mathbb{R}^1$, and the characteristic function of the exponential distribution with parameter $\alpha > 0$ is equal to

$$\alpha \int_0^{\infty} e^{i\lambda x}e^{-\alpha x}dx = \frac{\alpha}{\alpha - i\lambda}, \qquad \lambda \in \mathbb{R}^1,$$

so the result follows.   □

To continue the proof of ii), let, for each natural n, $\xi_1^n, \xi_2^n, \ldots$ be a sequence of independent random variables such that

$$\mathbb{P}(\xi_m^n = 1) = \alpha/n, \ \mathbb{P}(\xi_m^n = 0) = 1 - \alpha/n, m = 1, 2 \ldots,$$

and let $\pi^n(m)$ be the number of successes in the sequence $\xi_1^n, \xi_2^n, \ldots, \xi_m^n$. Define $m_l^n = [nt_l]$, where $[s]$ denotes the integer part of s. By the very definition, for each n, the random variables $\pi^n(m_1), \pi^n(m_2) - \pi^n(m_1), \ldots, \pi^n(m_k) - \pi^n(m_{k-1})$, are independent. By a straightforward generalisation of Proposition 5.2.5 the laws of

$$\pi^n(m_1), \pi^n(m_2) - \pi^n(m_1), \ldots, \pi^n(m_k) - \pi^n(m_{k-1})$$

converge weakly, as n tends to $+\infty$, to the law of

$$\Pi(t_1), \Pi(t_2) - \Pi(t_1), \ldots, \Pi(t_k) - \Pi(t_{k-1}),$$

and the required independence follows.   □

For a more direct proof of ii) see e.g. [4].

The following proposition gives another intuitive characterisation of exponential random variables.

PROPOSITION 5.2.6. *Assume that ξ is a positive random variable such that for all $t, s > 0$:*

$$\mathbb{P}(\xi > t + s|\xi > t) = \mathbb{P}(\xi > s).$$

Then there exists $\alpha > 0$ such that

(5.2.1) $$\mathbb{P}(\xi > t) = e^{-\alpha t}.$$

PROOF. Let $G(s) = \mathbb{P}(\alpha > s)$ for $s > 0$. Then G satisfies the functional equation: $G(t + s) = G(s)G(t)$ for all $t, s > 0$. The function G is right continuous and positive, so the functional equation has a unique solution of the required form.   □

5.3. – The Steinhaus construction

The construction of the Wiener and Poisson processes, presented in the previous section, took for granted existence of a sequence of independent random variables with prescribed distributions. Existence of such sequences follows from the Kolmogorov existence result. In this section we show that such sequences can be directly constructed if the Lebesgue measure on $[0, 1)$ is taken for granted. The construction goes back to H. Steinhaus [58].

THEOREM 5.3.1. *Let $\mu_1, \mu_2, \ldots$ be a sequence of probability measures on $\mathbb{R}^1$. There exists a sequence (ξ_n) of independent real-valued random variables defined on $([0, 1), \mathcal{B}([0, 1)), \mathbb{P})$, where $\mathbb{P}$ is the Lebesgue measure on $[0, 1)$, such that the distributions of ξ_n are exactly the measures μ_n, $n = 1, 2, \ldots$.*

PROOF. An arbitrary number $\omega \in [0, 1)$, with the exception of a countable set of numbers of the form $k/2^m$, $k, m = 0, 1, \ldots$, can be uniquely represented in the form

$$\omega = \sum_{n=1}^{+\infty} \frac{\epsilon_n}{2^n}, \qquad \text{where} \qquad \epsilon_n = 0 \text{ or } 1, \quad n = 1, 2, \ldots .$$

The sequence $(\epsilon_1, \epsilon_2, \ldots)$ is called the dyadic expansion of ω. Define

$$X_n(\omega) = \epsilon_n, \qquad \omega \in [0, 1), \ n = 1, 2, \ldots .$$

It is easy to see that if $\epsilon_i = 0$ or 1, $i = 1, 2, \ldots, n$, $n = 1, 2, \ldots$,

$$\{\omega \in [0, 1) : X_1(\omega) = \epsilon_1, \ldots, X_n(\omega) = \epsilon_n\} = \left[\frac{\epsilon_1}{2^1} + \ldots + \frac{\epsilon_n}{2^n}, \frac{\epsilon_1}{2^1} + \ldots + \frac{\epsilon_n}{2^n} + \frac{1}{2^n} \right).$$

This implies that

$$\mathbb{P}(X_1 = \epsilon_1, \ldots, X_n = \epsilon_n) = \frac{1}{2^n} = \prod_{i=1}^{n} \mathbb{P}(X_i = \epsilon_i)$$

and therefore the random variables (X_n) are independent.

Let $J_i = \{n_{i,j} : j = 1, 2, \ldots\}$, $i = 1, 2, \ldots$ be disjoint subsets of the set of natural numbers. Then the random variables

$$Z_i = \sum_{j=1}^{+\infty} \frac{X_{n_{i,j}}}{2^j}, \qquad i = 1, 2, \ldots$$

are independent. We show that they have uniform distribution on $[0, 1)$. Let, for instance, $i = 1$ and define

$$S_n = \sum_{j=1}^{n} \frac{X_{n_{1,j}}}{2^j}.$$

Then $\mathbb{P}(S_n = k/2^n) = 1/2^n$, $k = 0, 1, 2, \ldots, 2^n - 1$, and therefore, for $t \in [0, 1)$, $\mathbb{P}(S_n \le t) \to t$. On the other hand, $\mathbb{P}(S_n \le t) \to \mathbb{P}(Z_1 \le t)$. Thus $\mathbb{P}(Z_1 \le t) = t$.

Now let μ be a probability measure on $\mathbb{R}^1$ and let $F = F_\mu$ be its distribution function. Define the inverse F^{-1} of F by

$$F^{-1}(s) = \inf\{t : s \le F(t)\}, s \in [0, 1).$$

If Z has uniform distribution on $[0, 1)$ then the distribution of $F^{-1}(Z)$ is exactly μ. Indeed, from the definition of F^{-1}, for $s \in [0, 1)$ and $t \in (-\infty, +\infty)$, $s \le F(F^{-1}(s))$ and $F^{-1}(F(t)) \le t$. Therefore, $\{s : F^{-1}(s) \le t\} = [0, F(t)]$ and, consequently, $\mathbb{P}(\omega : F^{-1}(Z(\omega)) \le t) = \mathbb{P}(\omega : Z(\omega) \le F(t)) = F(t)$.

To finish the proof of the theorem, it is sufficient to remark that if F_1^{-1}, $F_2^{-1}, \ldots$ are functions, defined as above, corresponding to the measures μ_1, $\mu_2, \ldots$ and $Z_1, Z_2, \ldots$ are real-valued random variables uniformly distributed in $[0, 1)$ then the sequence $\xi_1 = F_1^{-1}(Z_1), \xi_2 = F_2^{-1}(Z_2), \ldots$ has all the properties required. $\square$

CHAPTER 6

Analytic approach to Lévy processes

For Lévy processes a complete characterisation of finite dimensional distributions is possible. It turns out that Lévy processes are far reaching generalisation of the Poisson process. The main result of the chapter is the so called Lévy-Khinchin formula. We derive also some basic properties of the corresponding transition semigroups both in $UC_b(\mathbb{R}^d)$ and in $L^p(\mathbb{R}^d)$. Subordination procedure is discussed as well.

6.1. – Infinitely divisible families

We recall that a family μ_t, $t \geq 0$ of probability measures on $\mathbb{R}^d$ is infinitely divisible if

1) $\mu_0 = \delta_{\{0\}}$,
2) $\mu_{t+s} = \mu_t * \mu_s$ for all $t, s \geq 0$,
3) $\mu_t \Rightarrow \delta_{\{0\}}$ as $t \to 0$.

The building blocks of the infinite divisible families are the *compound Poisson*, the *shift* and the *Gaussian* families, which we have already met in a slightly different context.

PROPOSITION 6.1.1. *Let ν be a probability measure on $\mathbb{R}^d$ and λ a positive constant. Then*

$$\mu_t = e^{-\lambda t} \sum_{n=0}^{\infty} \frac{(\lambda t)^n}{n!} \nu^{*n}, \; t > 0, \qquad \mu_0 = \delta_{\{0\}},$$

is an infinitely divisible family, called a compound Poisson family with parameters λ, ν. We are using the convention $\nu^{*0} = \delta_{\{0\}}$.

PROOF. It is clear that $\mu_t(\mathbb{R}^d) = 1$, $t \geq 0$.

$$\mu_t * \mu_s = e^{-\lambda(t+s)} \sum_{n,m} \frac{(\lambda t)^m}{m!} \frac{(\lambda s)^n}{n!} \nu^{*(n+m)}$$

$$= e^{-\lambda(t+s)} \sum_{k=0}^{\infty} \nu^{*k} \sum_{n+m=k} \frac{(\lambda t)^m (\lambda s)^n}{m! \, n!}$$

$$= e^{-\lambda(t+s)} \sum_{k=0}^{\infty} \nu^{*k} \frac{1}{k!} (\lambda t + \lambda s)^k$$

$$= \mu_{t+s}.$$

To check 3) fix a bounded continuous function ϕ. Then

$$\int_{\mathbb{R}^d} \phi(x) \mu_t(dx) = e^{-\lambda t} \phi(0) + e^{-\lambda t} \sum_{m=1}^{+\infty} \frac{(\lambda t)^m}{m!} \int_{\mathbb{R}^d} \phi(x) \nu^{*m}(dx).$$

As t tends to 0 the sum in the above expression tends to 0 as well and the required property follows. $\square$

If (μ_t) are compound Poisson distributions the corresponding process $X(t)$, $t \geq 0$, is called a *compound Poisson process*. It can be constructed as follows.

Let $Z_1, Z_2, \ldots$ be a sequence of independent random variables with identical laws equal to the measure ν. Let, in addition, $\Pi(t)$, $t \geq 0$, be a Poisson process, with parameter λ, independent of $Z_1, Z_2, \ldots$. Define

$$X(t) = 0 \qquad\qquad \text{if } \Pi(t) = 0$$
$$= Z_1 + \ldots + Z_k \quad \text{if } \Pi(t) = k.$$

Then $X(t)$, $t \geq 0$, is a compound Poisson process with parameters λ, ν. For instance if Γ is a Borel subset of $\mathbb{R}^d$ then

$$\mathbb{P}(X(t) \in \Gamma) = \mathbb{P}(X(t) \in \Gamma \text{ and } \Pi(t) = 0) + \sum_{k=1}^{\infty} \mathbb{P}(X(t) \in \Gamma \text{ and } \Pi(t) = k)$$

$$= e^{-\lambda t} \delta_{\{0\}}(\Gamma) + \sum_{k=1}^{\infty} \mathbb{P}(Z_1 + \ldots + Z_k \in \Gamma \text{ and } \Pi(t) = k)$$

$$= e^{-\lambda t} \delta_{\{0\}}(\Gamma) + \sum_{k=1}^{\infty} \mathbb{P}(Z_1 + \ldots + Z_k \in \Gamma) \mathbb{P}(\Pi(t) = k)$$

$$= e^{-\lambda t} \delta_{\{0\}}(\Gamma) + \sum_{k=1}^{\infty} \nu^{*k}(\Gamma) e^{-\lambda t} \frac{(\lambda t)^k}{k!} = e^{-\lambda t} \sum_{k=0}^{\infty} \frac{(\lambda t)^k}{k!} \nu^{*k}(\Gamma). \qquad \square$$

Fix $a \in \mathbb{R}^d$ and define

$$\mu_t = \delta_{\{ta\}} \qquad t \geq 0.$$

Then obviously (μ_t) is an infinitely divisible family and is called a *shift* family.

Gaussian infinitely divisible family on $\mathbb{R}^d$ is parameterised by a vector $a \in \mathbb{R}^d$ and a symmetric, non-negative, $d \times d$ matrix Q and given by the formula:

$$\mu_t = N(ta, tQ), \quad t \geq 0.$$

In particular, if Q is non-degenerate, then μ_t has a density and

$$\mu_t(dx) = \frac{1}{\sqrt{(2\pi t)^d \det Q}} e^{-\frac{1}{2t}<Q^{-1}(x-a),x-a>} dx, \quad t \geq 0.$$

6.2. – The Lévy-Khinchin formula

Let μ_t, $t \geq 0$, be the family of compound Poisson distributions with the intensity parameters $\lambda > 0$ and the jump measure ν. Then it is easy to calculate the characteristic function of μ_t, $t \geq 0$:

$$\widehat{\mu}_t(\xi) = \int_{\mathbb{R}^d} e^{i\langle \xi, x\rangle} \mu_t(dx) = e^{-\lambda t} \sum_{n=0}^{\infty} \frac{(\lambda t)^n}{n!} (\widehat{\nu}(\xi))^n$$

$$= e^{-\lambda t} e^{\lambda t \widehat{\nu}(\xi)} = e^{-t\psi(\xi)}, \quad t \geq 0, \xi \in \mathbb{R}^d,$$

where

$$\psi(\xi) = \lambda \int_{\mathbb{R}^d} (1 - e^{i\langle \xi, y\rangle}) \nu(dy).$$

The function ψ is called the exponent of the family (μ_t). For the shift family $\mu_t = \delta_{\{ta\}}$, we have

$$\widehat{\mu}_t(\xi) = \int_{\mathbb{R}^d} e^{i\langle \xi, x\rangle} \delta_{\{ta\}}(dx) = e^{-t\psi(\xi)}$$

where $\psi(\xi) = -i\langle a, \xi\rangle$. For the Gaussian family $\mu_t = N(0, tQ)$, $\widehat{\mu}_t(\xi) = e^{-t\frac{1}{2}\langle Q\xi, \xi\rangle}$, so the exponent ψ is of the form

$$\psi(\xi) = \frac{1}{2}\langle Q\xi, \xi\rangle, \quad \xi \in \mathbb{R}^d.$$

The following general characterisation, due to Lévy and Khinchin, takes place.

THEOREM 6.2.1. *For arbitrary infinitely divisible family (μ_t) on $\mathbb{R}^d$ there exist a vector $a \in \mathbb{R}^d$, a symmetric matrix $Q \geq 0$ and a nonnegative measure ν concentrated on $\mathbb{R}^d \backslash \{0\}$ satisfying*

$$(6.2.1) \qquad \int_{\mathbb{R}^d} (|y|^2 \wedge 1)\, \nu(dy) < \infty,$$

such that

$$(6.2.2) \qquad \widehat{\mu}_t(\xi) = e^{-t\psi(\xi)},$$

where

$$(6.2.3) \quad \psi(\xi) = i\langle a, \xi\rangle + \frac{1}{2}\langle Q\xi, \xi\rangle + \int_{\mathbb{R}^d} \left(1 - e^{i\langle \xi, y\rangle} + \frac{i\langle \xi, y\rangle}{1+|y|^2}\right) \nu(dy).$$

Conversely for given $a \in \mathbb{R}^d$, $Q \geq 0$ and a measure ν satisfying (6.2.1), formulae (6.2.2)-(6.2.3) define characteristic functions of an infinitely divisible family (μ_t).

Before going to the proof of the theorem we discuss some examples.

EXAMPLES.

1) If v is a finite measure and $\lambda = v(\mathbb{R}^d)$ then

$$\psi(\xi) = i\langle a, \xi\rangle + \frac{1}{2}\langle Q\xi, \xi\rangle + \lambda \int_{\mathbb{R}^d} (1 - e^{i\langle \xi, y\rangle})\frac{1}{\lambda} v(dy)$$

$$- i\left\langle \int_{\mathbb{R}^d} \frac{y}{1 + |y|^2} v(dy), \xi \right\rangle,$$

so ψ is a sum of the exponents introduced at the beginning.

2) Assume that $a = 0$, $Q = 0$ and the measure v on $\mathbb{R}^d$ is of the form:

$$(6.2.4) \qquad\qquad v(dx) = \frac{c}{|x|^{d+\alpha}} dx,$$

where $c > 0$, $\alpha \in (0, 2)$. Then the conditions of the Lévy-Khinchin theorem are satisfied and we have the following result.

PROPOSITION 6.2.2. *If $a = 0$, $Q = 0$ and v is given by (6.2.4) then*

$$\psi(\xi) = c_1 |\xi|^\alpha, \qquad \xi \in \mathbb{R}^d,$$

where c_1 is a positive constant.

PROOF. In the present case

$$\psi(\xi) = \int_{\mathbb{R}^d} (1 - \cos\langle \xi, x\rangle) \frac{c}{|x|^{d+\alpha}} dx.$$

It is clear that ψ is invariant under rotation around 0. Moreover if $\sigma > 0$ then

$$\psi(\sigma\xi) = \int_{\mathbb{R}^d} (1 - \cos\langle \xi, \sigma x\rangle) \frac{c}{|x|^{d+\alpha}} dx.$$

Changing coordinates $y = \sigma x$ one gets the result.   □

The infinitely divisible families described in the proposition are called α-stable, rotationally invariant families.

PROOF OF THE THEOREM. We show first that (6.2.1)-(6.2.3) define infinitely divisible laws. For $r > 0$ define

$$\psi_r(\xi) = \int_{|y|\geq r} \left(1 - e^{i\langle \xi, y\rangle} + \frac{i\langle \xi, y\rangle}{1 + |y|^2}\right) v(dy), \qquad \xi \in \mathbb{R}^d,$$

then ψ_r is the exponent of an infinitely divisible family. Note that if $|\xi|, |y| \leq R$ then for some $c > 0$:

$$|1 - (\cos\langle\xi, y\rangle)| \leq c|\langle\xi, y\rangle|^2 \leq c|y|^2,$$

$$\left|(\sin\langle\xi, y\rangle) - \frac{\langle\xi, y\rangle}{1 + |y|^2}\right| \leq c|\sin\langle\xi, y\rangle - \langle\xi, y\rangle| + \frac{|\langle\xi, y\rangle|\,|y^2|}{1 + |y|^2} \leq c|y|^2.$$

Consequently

$$\int_{|y|\leq r} \left|1 - e^{i\langle\xi, y\rangle} + \frac{i\langle\xi, y\rangle}{1 + |y|^2}\right| \frac{1}{|y|^2}|y|^2\, v(dy) \to 0,$$

as $r \to 0$, uniformly for ξ in a bounded set. Thus ψ_r converge uniformly on bounded sets to a continuous function ψ. This shows that the functions defined by (6.2.2) are characteristic functions of some measures. Since $\hat\mu_t \cdot \hat\mu_s = \hat\mu_{t+s}$, and $\hat\mu_t \to 1$ as $t \downarrow 0$, the family (μ_t) determined by (6.2.1)-(6.2.3) is infinitely divisible.

To prove that any infinitely divisible family is of the form (6.2.1)-(6.2.3) we will use a classical result in the theory of C_0-semigroups (P_t) on a Banach space X.

PROPOSITION 6.2.3. *If (P_t) is a C_0-semigroup on X then for some $\omega > 0$ and $M > 0$, $|P_t| \leq e^{\omega t} M$. Moreover if*

$$R_\lambda = \int_0^{+\infty} e^{-\lambda t} P_t dt, \qquad \lambda > \omega,$$

then

$$P_t f = \lim_{\lambda \to +\infty} e^{tA_\lambda} f,$$

where

$$A_\lambda = \lambda(\lambda R_\lambda - I). \qquad \lambda > \omega.$$

Let now (μ_t) be an infinitely divisible family. Define

$$P_t f(x) = \int_{\mathbb{R}^d} f(x + y)\,\mu_t(dy), \qquad f \in UC_b(\mathbb{R}^d),$$

where $UC_b(\mathbb{R}^d)$ denotes the space of all bounded, uniformly continuous functions on R^d, equipped with the supremum norm. Then, by Proposition 6.3.4, (P_t) is a C_0-semigroup on $UC_b(\mathbb{R}^d)$ and for $\lambda > 0$

$$\lambda R_\lambda f(x) = \lambda \int_0^{+\infty} e^{-\lambda t} \left[\int_{\mathbb{R}^d} f(x + y)\,\mu_t(dy)\right] dt = \int_{\mathbb{R}^d} f(x + y)\,\eta_\lambda(dy),$$

where η_λ is the probability measure $\eta_\lambda = \lambda \int_0^{+\infty} e^{-\lambda t}\mu_t\, dt$. (The question of measurability of $t \to \mu_t(\Gamma)$ arises here. Note that function $t \to \int f\mu_t(dy)$

is continuous for $f \in C_b(\mathbb{R}^d)$. Taking a sequence $f_n \downarrow \chi_\Gamma$ as in a previous proposition we conclude that measurability of $t \to \mu_t(\Gamma)$ holds for Γ closed. By Dynkin's $\pi - \lambda$ lemma it holds for Γ Borel.)

The semigroup

$$P_t^\lambda = e^{tA_\lambda}$$
$$= e^{-\lambda t} e^{t\lambda^2 R_\lambda} = e^{-\lambda t} \sum_{n=0}^\infty \frac{(t\lambda)^n}{n!} (\lambda R_\lambda)^n,$$

corresponds to the compound Poisson family with the intensity $\lambda > 0$ and the jump measure η_λ. It follows from the proposition that these compound Poisson families converge, for each $t > 0$, weakly to μ_t. Strictly speaking the proposition gives convergence for every function belonging to $UC_b(\mathbb{R}^d)$ but the generalisation to all functions from $C_b(\mathbb{R}^d)$ is immediate. More precisely we have that

$$(6.2.5) \qquad \widehat{\mu}_t(\xi) = \lim_{\lambda \to +\infty} e^{-t\lambda \int_{\mathbb{R}^d} \left(1 - e^{i\langle \xi, y \rangle}\right) \eta_\lambda(dy)}.$$

Set

$$(6.2.6) \qquad \psi_\lambda(\xi) = \int_{\mathbb{R}^d} (1 - e^{i\langle \xi, y \rangle}) \, v_\lambda(dy),$$

where

$$v_\lambda = \lambda \eta_\lambda$$

is a measure of total mass λ.

We show that (6.2.5) implies that

$$(6.2.7) \qquad \psi_\lambda(\xi) \to \psi(\xi), \qquad \text{as } \lambda \to +\infty,$$

where ψ is a continuous function. To see this note that $\widehat{\mu}_t \to 1$ uniformly on bounded sets as $t \to 0$. Therefore, for arbitrary $r > 0$, there exists $t > 0$ such that if $|\xi| \le r$ then,

$$|\widehat{\mu}_t(\xi) - 1| \le \frac{1}{4}.$$

Let (μ_t^λ) be the family corresponding to λ. Since $\widehat{\mu}_t^\lambda \to \widehat{\mu}_t$, uniformly on bounded sets as $\lambda \to +\infty$, therefore for some $\lambda_0 > 0$ and $\lambda > \lambda_0$

$$|e^{-t\psi_\lambda(\xi)} - 1| \le \frac{1}{2}.$$

Consequently

$$-t \operatorname{Re} \psi_\lambda(\xi) = \ln |\widehat{\mu}_t^\lambda(\xi)|,$$
$$-t \operatorname{Im} \psi_\lambda(\xi) = \operatorname{Arg} \widehat{\mu}_t^\lambda(\xi),$$

where Arg z for z such that $|z - 1| < \frac{1}{2}$ is a well defined continuous function. This shows existence and continuity of the limit ψ of ψ_λ. Thus

$$\widehat{\mu}_t = e^{-t\psi}$$

and we have to show that ψ is of the form (6.2.1)-(6.2.3).

Let us start with some heuristic considerations. In view of (6.2.6)-(6.2.7) it seems natural to investigate convergence of ν_λ. We can not expect convergence for $\lambda \to +\infty$, since $\nu_\lambda(\mathbb{R}^d) = \lambda$, so the total mass diverges to ∞. However, since $\nu_\lambda = \int_0^{+\infty} \lambda^2 e^{-\lambda t} \mu_t \, dt$ and $\mu_0 = \delta_{\{0\}}$, we see that ν_λ accumulates mass near 0. So it is reasonable to expect convergence of a measure of the form $K(y)\nu_\lambda(dy)$, where K is a function that tends to 0 as $|y| \to 0$. A reasonable attempt would be $K(y) = 1 \wedge |y|^2$ (compare (6.2.1)). For technical reasons, we will take a smoothed version of this function.

Now we come back to precise arguments. Let us fix $h > 0$ and define a function K:

$$K(y) = \frac{1}{(2h)^d} \int_{[-h,h]^d} (1 - \cos\langle \sigma, y \rangle) \, d\sigma = \left(1 - \prod_{k=1}^d \frac{\sin h y_k}{h y_k}\right),$$

$$y = (y_1, \dots, y_d) \in \mathbb{R}^d.$$

Then $K(y) > 0$ for $y \neq 0$, $\lim_{y \to 0} \frac{1}{|y|^2} K(y) = \frac{h^2}{3}$, $\lim_{|y| \to +\infty} K(y) = 1$. Denote

$$\widetilde{\nu}_\lambda(dy) = K(y)\nu_\lambda(dy), \qquad \lambda > 0.$$

We will show that

$$\widetilde{\nu}_\lambda \Rightarrow \widetilde{\nu}$$

where $\widetilde{\nu}$ is a finite measure on $\mathbb{R}^d$. To this end let us compute the Fourier transform $\widetilde{\psi}_\lambda$ of $\widetilde{\nu}_\lambda$. For $\xi \in \mathbb{R}^d$,

$$\int_{\mathbb{R}^d} e^{i\langle \xi, y \rangle} \widetilde{\nu}_\lambda(dy) = \int_{\mathbb{R}^d} e^{i\langle \xi, y \rangle} K(y) \, \nu_\lambda(dy)$$

$$= \int_{\mathbb{R}^d} e^{i\langle \xi, y \rangle} \left(1 - \frac{1}{(2h)^d} \int_{[-h,h]^d} e^{i\langle \sigma, y \rangle} \, d\sigma\right) \widetilde{\nu}_\lambda(dy)$$

$$= \frac{1}{(2h)^d} \int_{[-h,h]^d} \psi_\lambda(\xi + \sigma) \, d\sigma - \psi_\lambda(\xi),$$

where ψ_λ has been defined in (6.2.6). Since $\psi_\lambda \to \psi$ uniformly on bounded sets as $\lambda \to +\infty$, the Fourier transform of $\widetilde{\nu}_\lambda$ converges to the Fourier transform $\widetilde{\psi}$ of a bounded measure $\widetilde{\nu}$,

$$\widetilde{\psi}(\xi) = \int_{\mathbb{R}^d} e^{i\langle \xi, y \rangle} \widetilde{\nu}(dy).$$

LEMMA 6.2.4. *If a sequence of nonnegative measures (μ_n) converges weakly to a measure μ and a closed set Γ is such that $\mu(\partial\Gamma) = 0$ then the measures μ_n^Γ, equal to the the restrictions of μ_n to Γ, converge weakly to μ^Γ.*

PROOF. The lemma will be proved later, together with other elementary properties of the weak convergence. Below we will take $\Gamma = \{r \leq |y| \leq R\}$. $\square$

We fix numbers $0 < r < R < +\infty$ such that the $\tilde{\nu}$ measure of the spheres $\{|x| = r\}$, $\{|x| = R\}$ is zero. Then

$$\psi_\lambda(y) = -i \int_{|y|<R} \sin\langle \xi, y \rangle \, \nu_\lambda(dy) + \int_{r \leq |y| < R} (1 - \cos\langle \xi, y \rangle) \, \nu_\lambda(dy)$$

$$+ \int_{|y|<r} (1 - \cos\langle \xi, y \rangle) \, \nu_\lambda(dy) + \int_{R \leq |y|} (1 - e^{i\langle \xi, y \rangle}) \, \nu_\lambda(dy).$$

The limit of the last integral, as $\lambda \to +\infty$, is equal to

$$\int_{R \leq |y|} (1 - e^{i\langle \xi, y \rangle}) \, \nu(dy),$$

where $\nu(dy) = K^{-1}(y)\tilde{\nu}(dy)$. Therefore the limit of the first integral exists as well. Since

$$\int_{|y|<R} \sin\langle \xi, y \rangle \, \nu_\lambda(dy)$$

$$= \int_{|y|<R} \frac{\sin\langle \xi, y \rangle - \langle \xi, y \rangle}{|y|^2} (|y|^2 K^{-1}(y)) \tilde{\nu}_\lambda(dy) + \left\langle \xi, \int_{|y|<R} y \, \nu_\lambda(dy) \right\rangle$$

$$= I_1 + I_2,$$

and the functions $y \to \frac{\sin\langle \xi, y \rangle - \langle \xi, y \rangle}{|y|^2}$, $y \to |y|^2 K^{-1}(y)$ are continuous and bounded on $\mathbb{R}^d$ therefore the limit of I_1, as λ tends to ∞, exists and is equal to

$$\int_{|y|<R} (\sin\langle \xi, y \rangle - \langle \xi, y \rangle) \, \nu(dy).$$

Consequently the limit of I_2 exists as well and is of the form $\langle \xi, a \rangle$ for some $a \in \mathbb{R}^d$. In a similar way one can show that

$$\lim_{r \to 0} \left[\lim_{\lambda \to +\infty} \int_{r \leq |y| < R} (1 - \cos\langle \xi, y \rangle) \, \nu_\lambda(dy) \right] = \int_{0 < |y| < R} (1 - \cos\langle \xi, y \rangle) \, \nu(dy).$$

It remains to prove that for some $Q \geq 0$ and all $\xi \in \mathbb{R}^d$

$$\frac{1}{2}\langle Q\xi, \xi \rangle = \lim_{r \to 0} \left[\lim_{\lambda \to +\infty} \int_{|y|<r} (1 - \cos\langle \xi, y \rangle) \, \nu_\lambda(dy) \right].$$

However

$$\int_{|y|<r} (1 - \cos\langle \xi, y \rangle) \, \nu_\lambda(dy)$$

$$= \int_{|y|<r} \frac{1 - \cos\langle \xi, y \rangle - \frac{1}{2}\langle \xi, y \rangle^2}{|y|^2} |y|^2 K^{-1}(y) \tilde{\nu}_\lambda(dy) + \frac{1}{2} \int_{|y|<r} \langle \xi, y \rangle^2 \, \nu_\lambda(dy)$$

$$= J_\lambda(r, \xi) + \frac{1}{2}\langle Q_{r,\lambda}\xi, \xi \rangle.$$

It follows, again from the weak convergence of $\widetilde{\nu}_\lambda \Rightarrow \widetilde{\nu}$, that

$$J_\lambda(r, \xi)\underset{\lambda \to +\infty}{\longrightarrow} \int_{|y|<r} \left[1 - \cos\langle\xi, y\rangle - \frac{1}{2}\langle\xi, y\rangle^2\right] \nu(dy).$$

Consequently also $Q_{r,\lambda} \to Q_r$ as $\lambda \to +\infty$. Setting now $r \to 0$ and taking the limit along a sequence $r_n \to 0$ such that $\widetilde{\nu}(|x| = r_n) = 0$, we get the desired result. $\square$

6.3. – Infinitely divisible families and semigroups

6.3.1. – Infinitely divisible families on $[0, +\infty)$

Similarly as on $\mathbb{R}^d$, a family of probability measures $(\mu_t)_{t\geq 0}$ on $[0, +\infty)$ is called infinitely divisible if and only if

i) $\mu_0 = \delta_{\{0\}}$,
ii) $\mu_{t+s} = \mu_t * \mu_s$, $t, s \geq 0$,
iii) $\mu_t \Rightarrow \delta_{\{0\}}$ as $t \downarrow 0$.

The proper tool to study infinitely divisible families on $[0, +\infty)$ are Laplace transforms. Denote by $\widetilde{\mu}$ the Laplace transform of a measure on $[0, +\infty)$. Thus

$$\widetilde{\mu}_t(\xi) = \int_0^{+\infty} e^{-\xi x} \mu_t(dx), \qquad \xi \geq 0, t \geq 0.$$

If ν is a probability measure concentrated on $[0, +\infty)$ and $\lambda > 0$ a positive constant then

$$\mu_t = e^{-\lambda t} \sum_{n=0}^\infty \frac{(\lambda t)^n}{n!} \nu^{*n}, \qquad t \geq 0,$$

is an infinitely divisible family on $[0, +\infty)$. It is called the compound Poisson family. Note that, in this case,

$$\widetilde{\mu}_t(\xi) = e^{-t\psi(\xi)}, \qquad \text{where} \quad \psi(\xi) = \lambda \int_0^{+\infty} (1 - e^{-\xi x}) \nu(dx).$$

In a similar way as for $\mathbb{R}^d$ one can show that the following result holds.

THEOREM 6.3.1. *Laplace transforms $(\widetilde{\mu}_t)$ of an infinitely divisible family are of the form*:

(6.3.1) $$\widetilde{\mu}_t(\xi) = e^{-t\psi(\xi)}, \qquad \text{where} \quad \psi(\xi) = \lambda \int_0^{+\infty} (1 - e^{-\xi x}) \nu(dx),$$

where ν is a nonnegative measure on $(0, +\infty)$ such that

(6.3.2) $$\nu([1, +\infty)) < \infty, \qquad \int_0^1 x \, \nu(dx) < \infty.$$

Conversely, the formula (6.3.1) *defines an infinitely divisible family if ν satisfies* (6.3.2).

We also have

PROPOSITION 6.3.2. *Assume that* $v(dx) = \frac{c}{x^{1+\beta}} dx$, $\beta \in (0, 1)$, $c > 0$, *then*

$$\psi(\xi) = c_1 \xi^\beta, \qquad \xi \geq 0,$$

for some constant $c_1 > 0$.

PROOF. The same as of Proposition 6.2.2.

6.3.2. – Subordination

THEOREM 6.3.3. *Let* (μ_t), (v_t) *be infinitely divisible families on* $\mathbb{R}^d$ *and* $[0, +\infty)$ *with exponents* ψ *and* ϕ *respectively. Then the formula*

$$\sigma_t = \int_0^\infty \mu_s \, v_t(ds), \qquad t \geq 0,$$

defines an infinitely divisible family on $\mathbb{R}^d$ *with exponent* $\phi(\psi)$.

PROOF. We check first that $\sigma_t * \sigma_u = \sigma_{t+u}$. Indeed

$$\sigma_t * \sigma_u = \left[\int_0^\infty \mu_s \, v_t(ds) \right] * \left[\int_0^\infty \mu_r \, v_u(dr) \right]$$

$$= \int_0^\infty \int_0^\infty \mu_s * \mu_r \, v_t(ds) v_u(dr)$$

$$= \int_0^\infty \int_0^\infty \mu_{s+r} \, v_t(ds) v_u(dr)$$

$$= \int_0^\infty \mu_\sigma \, (v_t * v_u)(d\sigma) = \int_0^\infty \mu_\sigma \, v_{t+u}(d\sigma) = \sigma_{t+u}.$$

Moreover, by the definition of ψ:

$$\widehat{\sigma_t}(\xi) = \int_0^\infty \widehat{\mu_s}(\xi) \, v_t(ds) = \int_0^\infty e^{-s\psi(\xi)} \, v_t(ds).$$

The definition of ϕ therefore gives

$$\widehat{\sigma_t}(\xi) = e^{-t\phi(\psi(\xi))}, \qquad t \geq 0, \, \xi \in \mathbb{R}^d.$$

Note that ϕ is defined on the right half plane of $\mathbb{C}$ and $\psi(\xi) \geq 0$ for all $\xi \in \mathbb{R}^d$.   □

6.3.3. – Semigroups determined by infinitely divisible families

Let (P_t) be a semigroup of linear, continuous operators on the space $UC_b(\mathbb{R}^d)$ of continuous where $UC_b(\mathbb{R}^d)$ denotes the space of all bounded, uniformly continuous functions on $\mathbb{R}^d$, equipped with the supremum norm. If $P_t 1 = 1$, $t \geq 0$, and $P_t f \geq 0$ for $f \geq 0$ then (P_t) is called *Markovian*. For each $a \in \mathbb{R}^d$ define the translation $\tau_a f$ of a function f by the formula $\tau_a f(x) = f(x + a)$, $a \in \mathbb{R}^d$, $x \in \mathbb{R}^d$. The semigroup (P_t) is called *translation invariant* if for arbitrary $a \in \mathbb{R}^d$, $t \geq 0$ and $f \in UC_b(\mathbb{R}^d)$, $P_t(\tau_a f) = \tau_a(P_t f)$.

We set

$$(6.3.3) \qquad P_t f(x) = \int_{\mathbb{R}^d} f(x + y)\, \mu_t(dy),$$

where μ_t is an infinitely divisible family.

PROPOSITION 6.3.4. *If (P_t) is defined on $UC_b(\mathbb{R}^d)$ by the formula (6.3.3) then (P_t) is a C_0-semigroup on $UC_b(\mathbb{R}^d)$.*

PROOF.

i) Assume that $f \in UC_b(\mathbb{R}^d)$ and $t > 0$. Then

$$|P_t f(x) - P_t f(z)| = \left| \int_{\mathbb{R}^d} [f(x + y) - f(z + y)]\, \mu_t(dy) \right|$$

$$\leq \int_{\mathbb{R}^d} |f(x + y) - f(z + y)|\, \mu_t(dy).$$

For $\epsilon > 0$ there exists $\delta > 0$ such that if $|x - x'| < \delta$ then $|f(x) - f(x')| < \epsilon$. Thus if $|x - z| < \delta$ then also $|(x+y) - (z+y)| < \delta$ and $|f(x+y) - f(z+y)| < \epsilon$. This gives $|P_t f(x) - P_t f(y)| < \epsilon$.

ii) $|P_t f(x) - f(x)| = |\int_{\mathbb{R}^d} (f(x + y) - f(x))\, \mu_t(dy)|$. For $\epsilon > 0$ there exists $\delta > 0$ such that $|f(x + y) - f(x)| < \epsilon$ if $|y| < \delta$ and x arbitrary. Therefore

$$|P_t f(x) - f(x)| \leq \epsilon \int_{|y| \leq \delta} \mu_t(dy) + 2\|f\| \int_{|y| > \delta} \mu_t(dy)$$

$$\leq \epsilon + 2\|f\| \int_{|y| > \delta} \mu_t(dy).$$

Since $\mu_t \Rightarrow \delta_{\{0\}}$ as $t \downarrow 0$, $P_t f \to f$ uniformly. $\qquad \square$

In fact we have the following characterisation of infinite divisible families which proof we left as an exercise.

PROPOSITION 6.3.5. *A Markovian semigroup (P_t) on $UC_b(\mathbb{R}^d)$ is translation invariant if and only if*

$$P_t f(x) = \int_{\mathbb{R}^d} f(x + y)\, \mu_t(dy),$$

for some infinitely divisible family (μ_t).

PROPOSITION 6.3.6. *There exists a unique extension of the operators* (P_t) *given by* (6.3.3) *from uniformly continuous functions with bounded supports to the whole* $L^p(\mathbb{R}^d)$ *for any* $p \in [1, \infty)$. *The extended family is a* C_0-*semigroup on* $L^p(\mathbb{R}^d)$.

PROOF.

i) Extendability.
Note that

$$\|P_t f\|_{L^p}^p = \int_{\mathbb{R}^d} \left| \int_{\mathbb{R}^d} f(x + y) \, \mu_t(dy) \right|^p dx$$

$$\leq \int_{\mathbb{R}^d} \left[\int_{\mathbb{R}^d} |f(x+y)|^p \mu_t(dy) \right] dx = \int_{\mathbb{R}^d} \left[\int_{\mathbb{R}^d} |f(x+y)|^p dx \right] \mu_t(dy)$$

$$\leq \int_{\mathbb{R}^d} (\|f\|_{L^p}^p) \, \mu_t(dy) \leq \|f\|_{L^p}^p.$$

ii) Continuity of $P_t f$ as $t \downarrow 0$.
One can assume that f is continuous with bounded support, as such functions are dense in L^p and the norms of P_t are bounded by *i*). Let $f(x) = 0$ if $|x| \geq R$. Note that

$$\int_{\mathbb{R}^d} |P_t f(x) - f(x)|^p dx = \int_{\mathbb{R}^d} \left| \int_{\mathbb{R}^d} (f(x + y) - f(x)) \, \mu_t(dy) \right|^p dx$$

$$\leq \int_{\mathbb{R}^d} \left[\int_{\mathbb{R}^d} |f(x + y) - f(x)|^p dx \right] \mu_t(dy).$$

For any $r \in (0, R)$,

$$\int_{\mathbb{R}^d} |P_t f(x) - f(x)|^p dx \leq \int_{|y| \leq r} \left(\int_{|x| \leq 2R} |f(x + y) - f(x)|^p dx \right) \mu_t(dy)$$

$$+ \int_{|y| > r} \mu_t(dy)[2^p \|f\|_{L^p}^p] \leq I_1 + I_2.$$

By taking $r > 0$ sufficiently small the term I_1 is smaller than $\epsilon > 0$ for all $t > 0$. By taking $t > 0$ sufficiently small also the second term is smaller than ϵ. This completes the proof.   □

6.3.4. – Generators of (P_t) on $L^2(\mathbb{R}^d)$

We choose now $p = 2$ and extend P_t to the space $L^2_{\mathbb{C}}$ of all 2-summable complex functions. Denoting as before by $\hat{g}$ the Fourier transform (characteristic function) of g we have for f with compact support

$$\widehat{P_t f}(\lambda) = \int_{\mathbb{R}^d} e^{i\langle \lambda, x \rangle} [P_t f(x)] \, dx = \int_{\mathbb{R}^d} e^{i\langle \lambda, x \rangle} \left[\int_{\mathbb{R}^d} f(x + y) \mu_t(dy) \right] dx$$

$$= \int_{\mathbb{R}^d} e^{-i\langle \lambda, y \rangle} \left[\int_{\mathbb{R}^d} e^{i\langle \lambda, x + y \rangle} f(x + y) \, dx \right] \mu_t(dy)$$

$$= \hat{f}(\lambda) e^{-t\psi(-\lambda)}, \qquad t \geq 0, \, \lambda \in \mathbb{R}^d.$$

Consequently the Fourier image of P_t acts as a multiplication by $\exp(-t\psi(-\cdot))$. Therefore the domain of its generator A is of the form:

$$D(A) = \{g \in L_{\mathbb{C}}^2;\ \psi(-\cdot)\widehat{g}(\cdot) \in L_{\mathbb{C}}^2\}$$

and A is given by the formula:

$$\widehat{Ag}(\lambda) = -\psi(-\lambda)\widehat{g}(\lambda), \qquad \lambda \in \mathbb{R}^d.$$

In particular for the Wiener semigroup with $\psi(\lambda) = \frac{1}{2}|\lambda|^2,\ \lambda \in \mathbb{R}^d$:

$$\widehat{Ag}(\lambda) = -\frac{1}{2}|\lambda|^2\widehat{g}(\lambda),$$

therefore A can be identified as the Laplace operator Δ on $L_{\mathbb{C}}^2(\mathbb{R}^d)$.

In the case of stable processes on $\mathbb{R}^d$, with $\psi(\lambda) = |\lambda|^\alpha,\ \alpha \in (0, 2)$,

$$\widehat{Ag}(\lambda) = -|\lambda|^\alpha\widehat{g}(\lambda),$$

and A can be identified with a pseudodifferential operator with symbol $-|\lambda|^\alpha$. In fact in this case $A = -(-\Delta)^{\alpha/2}$ and one can write a formula for the semigroup generated by A in terms of the heat semigroup only, using the idea of subordination.

Let (P_t) denote the heat semigroup and (P_t^α) the semigroup corresponding to $\psi(\lambda) = |\lambda|^\alpha$, called the α-stable semigroup, $\alpha \in (0, 2)$. Note that

$$|\lambda|^\alpha = 2^{-\frac{\alpha}{2}}\left(\frac{1}{2}|\lambda|^2\right)^{\frac{\alpha}{2}} = \phi\left(\frac{1}{2}|\lambda|^2\right),$$

where $\phi(\xi) = 2^{-\frac{\alpha}{2}}\xi^{\frac{\alpha}{2}},\ \xi \geq 0$. But ϕ is the exponent of an $\frac{\alpha}{2}$-stable semigroup of measures μ_t^α on $[0, +\infty)$ and by the subordination formula

$$P_t^\alpha = \int_0^\infty P_s\,\mu_t^\alpha(ds).$$

CHAPTER 7

Representation of Lévy processes

We show here that Wiener's constructive approach can be applied to processes with discontinuous paths. In fact starting from a countable family of random variables we will construct an arbitrary Lévy process. The obtained expression will give an additional interpretation of the Lévy-Khinchin formula.

7.1. – Poissonian random measures

Let $(E, \mathcal{E})$ be a measurable space and ν a finite measure on E. We recall that a random variable N has Poisson distribution with parameter $c > 0$ if

$$\mathbb{P}(N = k) = e^{-c} \frac{c^k}{k!}, \qquad k = 1, 2, \ldots .$$

By direct calculation we obtain a formula for the Laplace transform of the Poisson distribution

$$\mathbb{E}\left(e^{-\alpha N}\right) = e^{-c(1-e^{-\alpha})}, \qquad \alpha \geq 0.$$

Let N have a Poisson distribution with parameter $c = \nu(E)$ and let $\xi_1, \xi_2, \ldots$ be a sequence of independent random variables with distribution $\frac{1}{\nu(E)} \nu$.

THEOREM 7.1.1. *For arbitrary $A \in \mathcal{E}$ define*

$$\pi(A) = \sum_{k=1}^{N} 1_A(\xi_k).$$

i) The random variable $\pi(A)$ has Poisson distribution with parameter $\nu(A)$.

ii) For disjoint sets $A_1, \ldots, A_m$ the corresponding random variables are independent and

$$\pi(A_1 \cup \ldots \cup A_m) = \pi(A_1) + \ldots + \pi(A_m), \qquad \mathbb{P} - \text{a.s.}$$

The family π is called a *Poissonian measure with intensity measure ν.*
PROOF.

i) Fix $\alpha > 0$ and $A \in \mathcal{E}$. Then, by independence,

$$\mathbb{E}\,(e^{-\alpha\pi(A)}) = \sum_{m=0}^{+\infty} \mathbb{E}\,(e^{-\alpha \sum_{k=1}^{m} 1_A(\xi_k)} 1_{N=m})$$

$$= \sum_{m=0}^{+\infty} \mathbb{P}(N = m) \prod_{k=1}^{m} \mathbb{E}\,(e^{-\alpha 1_A(\xi_k)})$$

$$= \sum_{m=0}^{+\infty} e^{-c} \frac{c^m}{m!} (\mathbb{E}\,(e^{-\alpha 1_A(\xi_1)}))^m.$$

However

$$\mathbb{E}\,(e^{-\alpha 1_A(\xi_1)}) = e^{-\alpha}\mathbb{P}(\xi_1 \in A) + 1 - \mathbb{P}(\xi_1 \in A) = \frac{\nu(A)}{c}(e^{-\alpha} - 1) + 1.$$

Finally

$$\mathbb{E}\,(e^{-\alpha\pi(A)}) = e^{-c}\, e^{c[\frac{\nu(A)}{c}(e^{-\alpha}-1)+1]} = e^{-c\nu(A)(1-e^{-\alpha})}, \qquad \alpha \geq 0.$$

ii) Assume that the sets $A_1, \ldots, A_m$ are disjoint. For arbitrary non-negative numbers $\alpha_1, \ldots, \alpha_m$:

$$\mathbb{E}(e^{-\sum_{j=1}^{m} \alpha_j \sum_{k=1}^{N} 1_A(\xi_k)}) = \sum_{l=0}^{+\infty} e^{-c} \frac{c^l}{l!} \mathbb{E}\,(e^{-\sum_{j=1}^{m} \alpha_j \sum_{k=1}^{l} 1_{A_j}(\xi_k)})$$

$$= \sum_{l=0}^{+\infty} e^{-c} \frac{c^l}{l!} \mathbb{E}\,(e^{-\sum_{k=1}^{l}(\sum_{j=1}^{m} \alpha_j 1_{A_j}(\xi_k))})$$

$$= e^{-c} \sum_{l=0}^{+\infty} \frac{1}{l!} [c\,\mathbb{E}\,(e^{-\sum_{j=1}^{m} \alpha_j 1_{A_j}(\xi_k)})]^l$$

$$= e^{-c}\, e^{c\,\mathbb{E}(-\sum_{j=1}^{m} \alpha_j 1_{A_j}(\xi_1))}.$$

Since

$$\mathbb{E}\left(-\sum_{j=1}^{m} \alpha_j 1_{A_j}(\xi_1)\right) = \sum_{j=1}^{m} e^{-\alpha_j}\mathbb{P}(\xi_1 \in A_j) + 1 - \mathbb{P}(\xi_1 \in A_j)$$

$$= \sum_{j=1}^{m} \frac{\nu(A_j)}{c}(e^{-\alpha_j} - 1) + 1,$$

therefore

$$\mathbb{E}(e^{-\sum_{j=1}^{m} \alpha_j \sum_{k=1}^{N} 1_A(\xi_k)}) = e^{-c} e^{c \left(\sum_{j=1}^{m} \frac{\nu(A_j)}{c} (e^{-\alpha_j} - 1) + 1 \right)}$$

$$= \prod_{j=1}^{m} e^{-\nu(A_j)(1 - e^{-\alpha_j})}$$

$$= \prod_{j=1}^{m} \mathbb{E}(e^{-\alpha_j \pi(A_j)}).$$

Since the multivariate Laplace transform of the vector $(\pi(A_1), \ldots, \pi(A_m))$ is equal to the product of the Laplace transform of $\pi(A_1), \ldots, \pi(A_m)$ therefore the required independence follows.   □

THEOREM 7.1.2. *Let ν be a σ-finite measure on E and $\nu_1, \nu_2, \ldots$ finite measures with disjoint supports such that $\nu = \sum_{m=1}^{+\infty} \nu_m$. Let $\pi^1, \pi^2, \ldots$ be Poissonian measures constructed for measures $\nu_1, \nu_2, \ldots$ with independent double system $(N_m, \xi_1^m, \xi_2^m, \ldots)$. Then the family*

$$\pi(A) = \sum_{m=1}^{+\infty} \pi^m(A), \qquad A \in \mathcal{E},$$

satisfies properties i) and ii) from the previous theorem.

PROOF. Since the proof is similar to the previous one it will be omitted.   □

It follows from the construction that the family $\pi(A)$, $A \in \mathcal{E}$, has the following structure. There exists a sequence (x_k) of E-valued random variables such that

$$\pi(A)(\omega) = \sum_{k=1}^{+\infty} \delta_{x_k(\omega)}(A), \qquad \omega \in \Omega, \; A \in \mathcal{E}.$$

Thus the family $\pi(A)$, $A \in \mathcal{E}$, can be identified with a random distribution of a countable number of points (x_k) and $\pi(A)$ is equal to the number of points in the set A. One can also integrate with respect to π. Assume that f is a measurable, real function defined on E and set

$$(7.1.1) \qquad \int_E f(x) \, \pi(dx) = \sum_{k=1}^{+\infty} f(x_k)$$

for all those f for which the series in (7.1.1) is convergent $\mathbb{P}$-a.s. One can also define the integral using the Lebesgue scheme: first for simple, non-negative functions $f = \sum_{j=1}^{M} \alpha_j 1_{A_j}$, $A_j \cap A_k = \emptyset$, $j \neq k$, by setting

$$\int_E f(x) \, \pi(dx) = \sum_{j=1}^{M} \alpha_j \, \pi(A_j),$$

then by monotone limits for all measurable non-negative f and finally, for arbitrary f, by splitting it into the difference of positive and negative parts.

We gather basic properties of the integral in the following theorem.

THEOREM 7.1.3.

i) For arbitrary non-negative measurable f

(7.1.2) $\quad \mathbb{E}\,(e^{-\alpha \int_E f(x)\,\pi(dx)}) = e^{-\int_E (1-e^{-\alpha f(x)})\,\nu(dx)}, \qquad \alpha \geq 0.$

ii) If $\int_E |f(x)|\,\pi(dx) < +\infty$ then

(7.1.3) $\quad \mathbb{E}\,(e^{i\lambda \int_E f(x)\,\pi(dx)}) = e^{-\int_E (1-e^{i\lambda f(x)})\,\nu(dx)}, \qquad \lambda \in \mathbb{R}^1.$

iii) If $\int_E |f(x)|\,\nu(dx) < +\infty$ then $\mathbb{E}\,|\int_E f(x)\,\pi(dx)| < +\infty$ and

$$\mathbb{E}\int_E f(x)\,\pi(dx) = \int_E f(x)\,\nu(dx).$$

iv) If $\int_E |f(x)|\,\nu(dx) < +\infty$ and $\int_E |f(x)|^2\,\nu(dx) < +\infty$ then

$$\mathbb{E}\left|\int_E f(x)\,\pi(dx) - \int_E f(x)\,\nu(dx)\right|^2 = \int_E |f(x)|^2\,\nu(dx).$$

v) If non-negative measurable functions $f_1, f_2, \ldots, f_M$ have disjoint supports then the random variables $\int_E f_1(x)\,\pi(dx), \ldots, \int_E f_M(x)\,\pi(dx)$ are independent.

PROOF. We prove only *i)* as the proofs of the other points go the same way.

If $f = \sum_{j=1}^M \alpha_j\,1_{A_j}$, $A_j \cap A_k = \emptyset$, $j \neq k$, then

$$\mathbb{E}(e^{-\alpha \int_E f(x)\,\pi(dx)}) = \prod_{j=1}^M \mathbb{E}\,(e^{-\alpha\alpha_j \pi(A_j)}) = \prod_{j=1}^M e^{-\nu(A_j)(1-e^{-\alpha\alpha_j})}$$

and the formula holds. By monotone passage to the limit one gets it for all non-negative measurable functions. $\qquad\square$

7.2. – Representation theorem

It is clear that the exponent ψ, see formula (6.2.3), of the characteristic function

$$\hat{\mu}_t(\lambda) = e^{-t\psi(\lambda)}, \qquad t \geq 0,\ \lambda \in \mathbb{R}^d,$$

of an infinitely divisible family can be written in equivalent way

(7.2.1) $\qquad \psi(\lambda) = i\langle a, \lambda\rangle + \langle Q\lambda, \lambda\rangle + \psi_0(\lambda),$

where

(7.2.2) $\quad \psi_0(\lambda) = \int_{|x|\geq 1} (1 - e^{i\langle\lambda,x\rangle})\,\nu(dx) + \int_{|x|<1} (1 - e^{i\langle\lambda,x\rangle} + i\langle\lambda, x\rangle)\,\nu(dx),$

where ν is a measure on $\mathbb{R}^d\backslash\{0\}$ satisfying

$$\int_{|x|<1} |x|^2\,\nu(dx) + \nu\{x : |x| \geq 1\} < +\infty.$$

The main result of the chapter is the following theorem. In the proof we follow basically [2].

THEOREM 7.2.1. *Assume that the exponent ψ_0 of the characteristic function $\widehat{\mu}_t$,*
$t \geq 0$, is of the form (7.2.2). Let π be the Poissonian measure on $[0, +\infty) \times (\mathbb{R}^d \setminus \{0\})$
with the intensity $l_1 \times \nu$, where l_1 denotes the Lebesgue measure.

i) The formula

(7.2.3)
$$X(t) = \int_0^t \int_{|x| \geq 1} x \, \pi(ds, dx)$$
$$+ \lim_{\epsilon \downarrow 0} \left[\int_0^t \int_{\epsilon \leq |x| < 1} x \, \pi(ds, dx) - \int_0^t \int_{\epsilon \leq |x| < 1} x \, \nu(ds, dx) \right]$$

defines a process with independent increments having exponent ψ_0. The limit in
(7.2.3) exists $\mathbb{P}$-almost surely uniformly on arbitrary time interval $[0, T]$ if ϵ tends
to 0 sufficiently fast. In particular (7.2.3) defines a càdlàg process.

ii) The process X given by (7.2.3) has trajectories with bounded variation if and
only if

$$\int_{|x| < 1} |x| \, \nu(dx) < +\infty.$$

PROOF. 1) Consider first the process

$$X_1(t) = \int_0^t \int_{|x| \geq 1} x \, \pi(ds, dx), \qquad t \geq 0.$$

If $0 = t_0 < t_1 < \ldots < t_M$ then

$$(7.2.4) \quad X_1(t_j) - X_1(t_{j-1}) = \int_{]t_{j-1}, t_j] \times \{x \, : \, |x| \geq 1\}} x \, \pi(ds, dx), \quad j = 1, 2, \ldots, M,$$

and since the sets $]t_{j-1}, t_j] \times \{x : |x| \geq 1\}$, $j = 1, 2, \ldots, M$, are disjoint there-
fore, by Theorem 7.1.2, the random variables (7.2.4) are independent. Moreover,
the sets $[0, T] \times \{x : |x| \geq 1\}$ contain only a finite number of points $\{(t_k, x_k)\}$
and therefore the trajectories of X_1 are càdlàg. By Theorem 7.1.3-*ii*),

$$\mathbb{E}(e^{i \langle \lambda, X_1(t) \rangle}) = \mathbb{E}(e^{i \langle \lambda, \int_0^t \int_{|x| \geq 1} x \, \pi(ds, dx) \rangle})$$
$$= e^{-\int_0^t \int_{|x| \geq 1} (1 - e^{i \langle \lambda, x \rangle}) \, \nu(dx) \, ds}$$
$$= e^{-t \int_{|x| \geq 1} (1 - e^{i \langle \lambda, x \rangle}) \, \nu(dx)}, \qquad \lambda \in \mathbb{R}^d.$$

Define, for $\epsilon \in (0, 1)$,

$$X_{2,\epsilon}(t) = \int_0^t \int_{\epsilon \leq |x| < 1} x \, \pi(ds, dx) - t \, \nu(\{x \, : \, \epsilon \leq |x| < 1\}), \qquad t \geq 0.$$

In a similar way one shows that the process $X_{2,\epsilon}$ has independent increments, is independent of X_1 and that

$$\mathbb{E}(e^{i\langle\lambda, X_{2,\epsilon}(t)\rangle}) = e^{-t\int_{\epsilon \leq |x| < 1}(1 - e^{i\langle\lambda, x\rangle} + i\langle\lambda, x\rangle)\,\nu(dx)}.$$

It is therefore enough to show that the sequence of processes $X_{2,\epsilon}$ converges $\mathbb{P}$-a.s. uniformly on any fixed time interval $[0, T]$. It follows from Theorem 7.1.3-iv) that the process $X_{2,\epsilon}$ is a martingale with finite second moment. In fact if $0 < \eta < \epsilon < 1$ then

$$\mathbb{E}\,|X_{2,\epsilon}(t) - X_{2,\eta}(t)|^2 = t\int_{\eta \leq |x| < \epsilon} |x|^2\,\nu(dx).$$

We have the following corollary from the theorem on Doob's inequalities.

LEMMA 7.2.2. *If $(Z(t))$ is a martingale with right continuous trajectories then for arbitrary $c > 0$*

$$\mathbb{P}\left(\sup_{0 \leq t \leq T} |Z(t)| \geq c\right) \leq \frac{1}{c}\,\mathbb{E}\,|Z(T)|.$$

PROOF. It is enough to prove that for arbitrary discrete time martingale (X_n) and arbitrary $c > 0$

$$\mathbb{P}\left(\sup_{1 \leq n \leq k} |X_n| \geq c\right) \leq \frac{1}{c}\,\mathbb{E}\,|X_k|.$$

Let us remark that

$$\mathbb{P}\left(\sup_{1 \leq n \leq k} |X_n| \geq c\right) \leq \mathbb{P}\left(\sup_{1 \leq n \leq k} X_n \geq c\right) + \mathbb{P}\left(\inf_{1 \leq n \leq k} X_n \leq -c\right)$$

and by inequalities 1) and 4) of the Doob theorem we have that

$$\mathbb{P}\left(\sup_{1 \leq n \leq k} |X_n| \geq c\right) \leq \frac{1}{c}\,(\mathbb{E}\,X_1 + 2\mathbb{E}\,X_k^-).$$

But $\mathbb{E}\,X_1 = \mathbb{E}\,X_k = \mathbb{E}\,X_k^+ - \mathbb{E}\,X_k^-$ and $\mathbb{E}\,|X_k| = \mathbb{E}\,X_k^+ + \mathbb{E}\,X_k^-$, so the result holds. $\square$

It follows from the lemma that for arbitrary $T > 0$

$$\mathbb{P}\left(\sup_{0 \le t \le T} |X_{2,\epsilon}(t) - X_{2,\eta}(t)| \ge c\right) \le \frac{1}{c} \mathbb{E}|X_{2,\epsilon}(T) - X_{2,\eta}(T)|$$

$$= \frac{1}{c}(\mathbb{E}|X_{2,\epsilon}(T) - X_{2,\eta}(T)|^2)^{1/2}$$

$$= \frac{\sqrt{T}}{c}\left(\int_{\eta \le |x| < \epsilon} |x|^2 \, \nu(dx)\right)^{1/2} \to 0 \qquad \text{as } \eta, \epsilon \to 0.$$

Consequently there exists a process $X_2(t)$, $t \in [0, T]$, and a sequence $\epsilon_n \downarrow 0$ such that

$$\mathbb{P}\left(\lim_{n \to +\infty} \sup_{0 \le t \le T} |X_{2,\epsilon_n}(t) - X_2(t)| = 0\right) = 1.$$

This completes the proof of i).

ii) Let $\Delta X_t = X(t) - X(t-)$, $t \in (0, T]$. Then

$$\mathbb{E}\left(e^{-\sum_{0 < t \le T} |\Delta X_t|}\right) = \mathbb{E}\left(e^{-\int_0^{+\infty} \int_{\mathbb{R}^d} f(t,x) \, \pi(dt,dx)}\right),$$

where $f(t, x) = 1_{[0,T]}(t)|x|$. Consequently

$$\mathbb{E}\left(e^{-\sum_{0 < t \le T} |\Delta X_t|}\right) = e^{-T \int_{\mathbb{R}^d} (1 - e^{-|x|}) \, \nu(dx)}.$$

Thus if $\int_{\mathbb{R}^d}(1 - e^{-|x|}) \, \nu(dx) = +\infty$ then $\mathbb{P}(\sum_{0 < t \le T} |\Delta X_t| = +\infty) = 1$. If $\int_{\mathbb{R}^d}(1 - e^{-|x|}) \, \nu(dx) < +\infty$, equivalently $\int_{\mathbb{R}^d} |x| \, \nu(dx) < +\infty$, then

$$\mathbb{E} \sum_{0 < t \le T} |\Delta X_t| 1_{|X(t)| \le M} = T \int_{|x| \le M} |x| \, \nu(dx) < +\infty,$$

so X is of bounded variation. $\qquad \square$

CHAPTER 8

Stochastic integration and Markov processes

Wiener's idea to construct stochastic processes starting from simple probabilistic objects has been extended to all Markov processes by K. Ito in his seminal paper *On stochastic differential equations* [32]. His starting point was a Wiener process and a Poissonian random measure. He first developed a stochastic integration theory with respect to Wiener processes and Poissonian random measures. Then he introduced stochastic equations and showed that Markov processes can be obtained as their solutions. In this chapter we describe the Ito construction. We follow our preprint ([64], Lecture 1).

8.1. – Constructing Markov chains

Markov chains are discrete time analogues of Markov processes. We present here a construction of Markov chains as solutions of discrete time stochastic equations to motivate better Ito'approach.

Let $P(x, \Gamma)$, $x \in E$, $\Gamma \in \mathcal{E}$, be a transition kernel on a measurable space $(E, \mathcal{E})$. Thus for each x, $P(x, \cdot)$ is a probability measure on E and for each $\Gamma \in \mathcal{E}$, $P(\cdot, \Gamma)$ is an $\mathcal{E}$-measurable function. By P^n, $n = 0, 1, \ldots$, we define the iterated kernels

$$P^{n+1}(x, \Gamma) = \int_E P(x, dy) \, P^n(y, \Gamma), \qquad x \in E, \, \Gamma \in \mathcal{E}.$$

As in continuous time we define by induction, for any sequence of non-negative integers $n_1 < n_2 < \ldots < n_d$, the probabilities of visiting the sets $\Gamma_1, \ldots, \Gamma_d$ at moments $n_1, \ldots, n_d$ starting from x as the functions $P^{n_1, \ldots, n_d} : E \to \mathcal{P}(E^d)$:

$$P^{n_1, \ldots, n_d}(x, \Gamma_1, \ldots, \Gamma_d) = \int_{\Gamma_1} P^{n_1}(x, dx_1) \, P^{n_2 - n_1, \ldots, n_d - n_1}(x_1, \Gamma_2, \ldots, \Gamma_d).$$

A Markov chain X with the transition kernel P and starting from $x \in E$ is a sequence X_n, $n = 0, 1, \ldots$, of E-valued random variables on a probability space, such that

1) $X_0 = x$, $\mathbb{P}$-a.s.
2) $\mathbb{P}(\{\omega : X_{n_j}(\omega) \in \Gamma_j, \ j = 1, \ldots, d\}) = P^{n_1, \ldots, n_d}(x, \Gamma_1, \ldots, \Gamma_d)$.

We have the following result.

THEOREM 8.1.1. *Let E be a Polish space and $\mathcal{E} = \mathcal{B}(E)$ the family of its Borel sets. Then for any transition kernel P there exists a measurable mapping $F : E \times [0, 1) \to E$ such that, for an arbitrary sequence of independent random variables $\xi_1, \xi_2, \ldots$ with uniform distribution on $[0, 1)$, the inductively defined sequence X_n:*

$$(8.1.1) \qquad X_0 = x, \qquad X_{n+1} = F(X_n, \xi_{n+1}), \qquad n = 0, 1, \ldots$$

is a Markov chain with the transition kernel P.

PROOF. We construct F in the case when E is a countable set and $E = \mathbb{R}^1$. Let $E = \{1, 2, \ldots\} = \mathbb{N}$, $p_{n,m} = P(n, \{m\})$, $n, m \in \mathbb{N}$. We define

$$F(n, u) = m \qquad \text{for} \qquad u \in [p_{n,1} + \ldots + p_{n,m-1}, p_{n,1} + \ldots + p_{n,m}).$$

The required measurability is obvious and if ξ has uniform distribution on $[0, 1)$ then

$$\mathbb{P}(\{\omega : F(n, \xi(\omega)) = m\}) = P(n, \{m\}),$$

as required.

If $E = \mathbb{R}^1$ then we define first a measurable function

$$G(x, v) = P(x, (-\infty, v]), \qquad x \in \mathbb{R}^1, \ v \in \mathbb{R}^1,$$

and set

$$F(x, u) = \inf\{v : u \le G(x, v)\}, \qquad u \in [0, 1),$$

as in the proof of the Steinhaus theorem it is therefore clear that if ξ has uniform distribution on $[0, 1)$ then

$$(8.1.2) \qquad \mathbb{P}(\{\omega : F(x, \xi(\omega)) \in \Gamma\}) = P(x, \Gamma), \qquad x \in \mathbb{R}^1, \ \Gamma \in \mathcal{B}(\mathbb{R}^1).$$

A direct construction of F is possible also in the general case but we will limit ourselves to recalling a general result of Kuratowski saying (in particular) that if E is Polish then there exists a one-to-one measurable mapping S from E either on a finite set $\{1, 2, \ldots, N\}$ or on $\mathbb{N}$ or on $\mathbb{R}^1$ having measurable inverse. It is therefore clear that a function F can be constructed also for the Polish spaces.

The fact that (X_n) is a Markov chain with the transition kernel P follows from the independence of $\xi_1, \xi_2, \ldots$ by an induction argument. $\square$

REMARK. If the state space E is linear equation (8.1.1) can be rewritten as a difference, equation

(8.1.3) $\Delta X_n = H(X_n)\Delta Z_n$, $X_0 = x$,

where $\Delta X_n = X_{n+1} - X_n$, $\Delta Z_n = Z_{n+1} - Z_n$ and

(8.1.4) $Z_n = (n, \delta_{\xi_1} + \ldots + \delta_{\xi_n})$, $Z_0 = (0, 0)$,

(8.1.5) $H(x)(\alpha, \nu) = -\alpha x + \displaystyle\int_{[0,1]} F(x, v)\nu(dv)$.

In (8.1.5), α is a real number an ν a finite measure on $[0, 1]$. Note that the sequence (Z_n) has independent increments.

Thus any Markov chain can be generated from a sequence with independent increment and Ito showed that a similar result is true for general Markov processes in continuous time (see Section 8.4.3).

8.2. – The Courrège theorem

Let P^t be a transition function on $(\mathbb{R}^d, \mathcal{B}(\mathbb{R}^d))$. For each bounded measurable function ϕ on $\mathbb{R}^d$ we set

$$P^t\phi(x) = \int_{\mathbb{R}^d} P^t(x, dy)\,\phi(y), \qquad x \in \mathbb{R}^d.$$

We say that a transition function P^t is Feller if
1) $P^t\phi \in C_0(\mathbb{R}^d)$, for $\phi \in C_0(\mathbb{R}^d)$.
2) $P^t\phi \to \phi$ as $t \downarrow 0$, uniformly, for all $\phi \in C_0(\mathbb{R}^d)$.

The following theorem is due to Ph. Courrège [11]. In its formulation $C_0^\infty(\mathbb{R}^d)$ stands for the space of infinitely differentiable functions vanishing at infinity with all their derivatives and $L_+(\mathbb{R}^d, \mathbb{R}^d)$ for the set of symmetric, non-negative, $d \times d$ matrices.

THEOREM 8.2.1. *Let P^t be a Feller transition function such that for all $\phi \in C_0^\infty(\mathbb{R}^d)$ and all $x \in \mathbb{R}^d$, the function $P^t\phi(x)$, of the parameter $t \geq 0$, is differentiable. Then there exist transformations $F : \mathbb{R}^d \to \mathbb{R}^d$, $Q : \mathbb{R}^d \to L_+(\mathbb{R}^d, \mathbb{R}^d)$ and a family $\nu(x, \cdot)$, $x \in \mathbb{R}^d$, of non-negative measures concentrated on $\mathbb{R}^d \backslash \{0\}$ and satifying*

$$\int_{\mathbb{R}^d} (|y|^2 \wedge 1)\, \nu(x, dy) < +\infty, \qquad x \in \mathbb{R}^d,$$

such that for all $x \in \mathbb{R}^d$, $\phi \in C_0^\infty(\mathbb{R}^d)$

$$\lim_{t \to 0} \frac{P^t\phi(x) - \phi(x)}{t} = A\phi(x)$$

where

$$A\phi(x) = \langle F(x), D\phi(x) \rangle + \frac{1}{2} \text{ Trace } [Q(x)D^2\phi(x)]$$

$$+ \int_{\mathbb{R}^d} \left(\phi(x+y) - \phi(x) - \frac{\langle y, D\phi(x) \rangle}{|y|^2 + 1} \right) \nu(x, dy).$$

The functions F, Q and the measure ν are called *characteristics* of P^t or of the process. They have interpretations as *local drift*, *local diffusion* and *local jump measure* of the corresponding Markov process. The operator A is, the so called, *generator* of P^t or generator of the corresponding Markov process.

We will not prove this theorem but will make only some comments.

Let us notice that Proposition 6.2.3 states that the transition function (P^t) is a limit of transition functions (P_λ^t) of the following form:

$$P_\lambda^t \phi = e^{t\lambda(\lambda R_\lambda - I)} \phi = e^{-\lambda t} \sum_{k=0}^{+\infty} \frac{(\lambda t)^k}{k!} \widetilde{P}_\lambda^k \phi$$

where $\widetilde{P}_\lambda$ is a transition kernel defined for all bounded measurable ϕ:

$$\widetilde{P}_\lambda \phi = \lambda \int_0^{+\infty} e^{-\lambda t} P^t \phi \, dt.$$

The operator $\widetilde{P}_\lambda$ is bounded on $C_0(\mathbb{R}^d)$. Notice that for the approximating transition semigroup (P_λ^t), called also *compound Poissonian semigroup*, the operator A from the Courrège theorem is of the form

$$A_\lambda \phi(x) = \lambda \int_{\mathbb{R}^d} (\phi(x+y) - \phi(x)) \, \widetilde{P}_\lambda(x, dy).$$

Thus for the compound Poissonian semigroup, the drift and diffusion vanish and the jump measure is equal to $\lambda \widetilde{P}_\lambda(x, \cdot)$, $x \in \mathbb{R}^d$. As for Lévy's processes the structure of the limiting semigroup (P^t) is more complex. From the proof of the Lévy-Khinchin formula we could deduce the following result.

THEOREM 8.2.2. *Let (P^t) be a transition function of the form*

$$P^t \phi(x) = \int_{\mathbb{R}^d} \phi(x+y) \, \mu_t(dy),$$

where (μ_t) is an infinitely divisible family of measures with the representation (6.2.2)-(6.2.3). Then (P^t) satisfies the assumptions of the Courrège theorem and for $\phi \in C_0^\infty(\mathbb{R}^d)$:

$$A\phi(x) = \langle a, D\phi(x) \rangle + \frac{1}{2} \text{ Trace } [QD^2\phi(x)]$$

$$+ \int_{\mathbb{R}^d} \left(\phi(x+y) - \phi(x) - \frac{\langle y, D\phi(x) \rangle}{|y|^2 + 1} \right) \nu(dy).$$

The operator

$$A_0\phi(x) = \langle a, D\phi(x)\rangle + \frac{1}{2} \text{ Trace } [QD^2\phi(x)]$$

corresponds to the process:

$$X(t) = a\,t + Q^{1/2}W(t), \qquad t \geq 0,$$

where W is a Wiener process on $\mathbb{R}^d$ with identity covariance. The Lévy process with the integral part of A was constructed in an earlier chapter with a use of Poissonian random measures.

8.3. – Diffusion with additive noise

In the already quoted paper [32] Ito constructed Markov processes for a large family of characteristics $(F(.), Q(.), \nu(.))$. In the simpler case when $\nu = 0$ and the diffusion matrix is constant the construction does not require any stochastic calculus as we will show now.

Let $R = (r_{kl})$ be a non-negative $m \times m$ matrix, W an m-dimensional continuous Wiener process with components $W_1, W_2, \ldots, W_m$ such that

$$\mathbb{E}\,W_k(t) = 0, \quad \mathbb{E}\,(W_k(t)W_l(t)) = (t \wedge s)r_{kl}, \quad t, s \geq 0, \ k, l = 1, \ldots m.$$

Let, in addition, $F(x) = (F_1(x), \ldots, F_d(x))$, $x \in \mathbb{R}^d$, and B be a $d \times m$ matrix. Ito's equation, formally written as

$$dX(t) = F(X(t))\,dt + B\,dW(t), \qquad X(0) = x,$$

should be understood as an integral equation

(8.3.1) $\qquad X(t, \omega) = x + \int_0^t F(X(s, \omega))\,ds + B\,W(t, \omega), \qquad t \geq 0.$

We have the following result.

THEOREM 8.3.1. *Assume that the mapping F satisfies a Lipschitz condition. Then for all $\omega \in \Omega$, $x \in \mathbb{R}^d$, the equation (8.3.1) has a unique solution X^x. It is a Markov process starting from x with respect to a transition function having the generator*

(8.3.2) $\qquad A\phi(x) = \langle F(x), D\phi(x)\rangle + \frac{1}{2} \text{ Trace } [QD^2\phi(x)], \qquad \phi \in C_0^2(\mathbb{R}^d)$

where $Q = BRB^*$.

The existence of the solution to (8.3.2) can be proved by a fixed point method in the space of continuous functions $C([0, T], \mathbb{R}^d)$, for any $T > 0$. Note also that the form of A is obvious if $Q = 0$.

We close this section by presenting examples.

The stochastic harmonic oscillator equation

$$(8.3.3) \qquad \frac{d^2 X}{dt^2} = -\alpha X + \frac{dW}{dt}, \qquad X(0) = x_0, \qquad \frac{dX}{dt}(0) = v_0,$$

can be written in a vector form:

$$(8.3.4) \qquad d \begin{pmatrix} X(t) \\ v(t) \end{pmatrix} = \begin{pmatrix} 0 & 1 \\ -\alpha & 0 \end{pmatrix} \begin{pmatrix} X(t) \\ v(t) \end{pmatrix} dt + \begin{pmatrix} 0 \\ 1 \end{pmatrix} dW(t)$$

where W is a 1-dimensional Wiener process and $\alpha \geq 0$. It has the following solution:

$$X(t) = (\cos \sqrt{\alpha} t) x_0 + \frac{1}{\sqrt{\alpha}} (\sin \sqrt{\alpha} t) v_0 + \frac{1}{\sqrt{\alpha}} \int_0^t \sin(\sqrt{\alpha}(t - s)) \, dW(s)$$

$$v(t) = -\sqrt{\alpha} (\sin \sqrt{\alpha} t) + (\cos \sqrt{\alpha} t) v_0 + \int_0^t \cos(\sqrt{\alpha}(t - s)) \, dW(s).$$

The integrals with respect to W are the usual Stieltjes integrals.

The equation (8.3.4) is a special case of the linear equation

$$(8.3.5) \qquad dX = AX \, dt + B \, dW, \qquad X(0) = x \in \mathbb{R}^d,$$

where A and B are respectively $d \times d$ and $d \times m$ matrices and W is an m-dimensional Wiener process with covariance R. Again the equation (8.3.5) should be understood in the integral form

$$X(t) = x + \int_0^t AX(s) \, ds + B \, W(t), \qquad t \geq 0.$$

The solution to (8.3.5) can be written explicitly

$$(8.3.6) \quad X(t) = e^{At} x + \int_0^t e^{A(t-s)} B \, dW(s) = e^{At} x + B \, W(t) + \int_0^t A e^{A(t-s)} B W(s) \, ds,$$

and this can be easily checked by substitution. Note also that

$$\mathcal{L}(X(t)) = N(e^{At} x, Q_t),$$

where

$$Q_t = \int_0^t e^{As} B B^* e^{A^* s} \, ds$$

and

$$P^t \phi(x) = \int_{\mathbb{R}^d} \phi(e^{At} x + y) \, N(0, Q_t)(dy), \qquad t \geq 0.$$

Assuming that ϕ has continuous and bounded Fréchet derivatives up to order 2 one checks, by applying Taylor's formula to ϕ, that

$$A\phi(x) = \lim_{t \downarrow 0} \frac{P^t \phi(x) - \phi(x)}{t} = \langle Ax, D\phi(x) \rangle + \frac{1}{2} \, \text{Trace} \, [Q D^2 \phi(x)], \qquad x \in \mathbb{R}^d,$$

where $Q = BRB^*$, which confirms the formula (8.3.2).

8.4. – Stochastic integrals

We pass now to the general case when the diffusion Q may depend on the state and the jump measure does not vanish. To introduce stochastic equations which determine Markov processes with given characteristics we need stochastic integrals with respect to a Wiener process W on $\mathbb{R}^m$ and with respect to a Poissonian random measure π on $[0, +\infty) \times \mathbb{R}^m$. Let W be a $\mathbb{R}^m$-valued Wiener process with covariance $R = (r_{kl})$ and π a Poissonian measure with the intensity μ on $[0, +\infty) \times \mathbb{R}^m$. We can assume that W and π are defined on a probability space $(\Omega, \mathcal{F}, \mathbb{P})$ equipped with a filtration $(\mathcal{F}_t)$ satisfying the usual conditions and that they satisfy the following conditions:

1) W has continuous paths, $W(0) = 0$.
2) $W(t)$ is $\mathcal{F}_t$-measurable and $\sigma(W(t) - W(s))$ is independent of $\mathcal{F}_s$ for all $t \geq s \geq 0$.
3) $\pi(\Gamma)$ is $\mathcal{F}_t$-measurable for each Borel set $\Gamma \subset [0, t] \times \mathbb{R}^m$ and $\sigma(\pi(\Delta))$ is independent of $\mathcal{F}_t$ for each Borel set $\Delta \subset (t, +\infty) \times \mathbb{R}^m$.
4) The σ-fields generated by W and π are independent.

8.4.1. – Integration with respect to a Wiener process W

The aim is to define

$$\int_0^t \Phi(s) \, dW(s), \qquad t \geq 0,$$

where Φ is a stochastic process whose values are in the space $L(\mathbb{R}^m, \mathbb{R}^d)$ of linear operators from $\mathbb{R}^m$ to $\mathbb{R}^d$. One starts with the simple integrands which are of the form

$$(8.4.1) \qquad \Phi(s) = \sum_{j=1}^k 1_{(t_j, t_{j+1}]}(s) \, \Phi_j$$

where $0 \leq t_1 < t_2 < \ldots < t_{k+1}$ is any finite sequence and Φ_j is an $L(\mathbb{R}^m, \mathbb{R}^d)$-valued random variable, $\mathcal{F}_{t_j}$-measurable, such that $\mathbb{E}|\Phi_j|^2 < +\infty$. For Φ of the form (8.4.1) one sets

$$\int_0^t \Phi(s) \, dW(s) = \sum_{j=1}^k \Phi_j \, (W(t \wedge t_{j+1}) - W(t \wedge t_j)).$$

It is clear that the stochastic integral is a square-integrable martingale with continuous trajectories. We have the following fundamental *isometric formula*:

PROPOSITION 8.4.1. *For an arbitrary simple process Φ and $t \geq 0$ one has*

$$(8.4.2) \qquad \mathbb{E} \left| \int_0^t \Phi(s) \, dW(s) \right|^2 = \mathbb{E} \int_0^t \|\Phi(s) R^{1/2}\|_{HS}^2 \, ds,$$

where, for an arbitrary operator $C \in L(\mathbb{R}^m, \mathbb{R}^d)$, $\|C\|_{HS} = (\sum_{j=1}^m |Ce_j|^2)^{1/2}$ and $e_1, \ldots e_m$ is an orthonormal basis of $\mathbb{R}^m$.

Proof. Let Φ_j be represented as a matrix $(\Phi_j(k,l))$ and let $W_1(t),\ldots,W_m(t)$ be the components of $W(t)$. Then for $t \geq s$ and Φ, $\mathcal{F}_s$-measurable and $W(t) - W(s)$ independent of $\mathcal{F}_s$ we have

$$\mathbb{E}\,|\Phi_j(W(t) - W(s))|^2$$
$$= \sum_{k=1}^{d} \mathbb{E}\left|\sum_{l=1}^{m} \Phi_j(k,l)(W_l(t) - W_l(s))\right|^2$$
$$= \sum_{k=1}^{d} \sum_{l_1=1}^{m} \sum_{l_2=1}^{m} \mathbb{E}\left[\Phi_j(k,l_1)\Phi_j(k,l_2)(W_{l_1}(t) - W_{l_1}(s))(W_{l_2}(t) - W_{l_2}(s))\right].$$

But

$$\mathbb{E}\,[\Phi_j(k,l_1)\Phi_j(k,l_2)(W_{l_1}(t) - W_{l_1}(s))(W_{l_2}(t) - W_{l_2}(s))\,|\mathcal{F}_s]$$
$$= \Phi_j(k,l_1)\Phi_j(k,l_2)\,\mathbb{E}\,[(W_{l_1}(t) - W_{l_1}(s))(W_{l_2}(t) - W_{l_2}(s))]$$
$$= \Phi_j(k,l_1)\Phi_j(k,l_2)\,r_{l_1,l_2}\,(t - s),$$

and therefore

$$\mathbb{E}\,|\Phi_j(W(t) - W(s))|^2 = (t - s)\ \text{Trace}\ [\Phi_j R \Phi_j^*] = (t - s)\,\|\Phi_j R^{1/2}\|_{HS}^2.$$

Thus the formula (8.4.2) follows for $k = 1$. The general case can be obtained by induction and proper conditioning. $\square$

The isometric formula is of fundamental importance. It allows to extend the integral to processes Φ which are limits of simple processes in the norms

$$\mathbb{E}\int_0^T \|\Phi(s)R^{1/2}\|_{HS}^2\,ds$$

for each T.

These processes are again $(\mathcal{F}_t)$-adapted, the integral is a square integrable martingale and the formula (8.4.2) holds. The integration can be extended to even larger classes of integrands which are $(\mathcal{F}_t)$-adapted and for which

$$\mathbb{P}\left(\int_0^t \|\Phi(s)R^{1/2}\|_{HS}^2\,ds < +\infty, \quad t \geq 0\right) = 1.$$

8.4.2. – Integration with respect to a Poissonian measure π

Let π be a Poissonian measure satifying 3) with the intensity μ. We will denote by $\mathcal{E}_0$ the family of all sets Γ from $\mathcal{E} = \mathcal{B}([0, +\infty) \times \mathbb{R}^d)$ for which $\mu(\Gamma) < +\infty$. The co called *compensated Poissonian measure* $\check{\pi}$ is given by the formula

$$\check{\pi}(\Gamma) = \pi(\Gamma) - \mu(\Gamma), \qquad \Gamma \in \mathcal{E}_0.$$

Our aim is to define first stochastic integrals:

$$(8.4.3) \qquad \int_0^t \int_{\mathbb{R}^d} \phi(s, x)\, \check{\pi}(ds, dx),$$

for a class of stochastic random fields ϕ. Let us recall that stochastic integrals of the type (8.4.3), with deterministic ϕ, were introduced in the section on random measures. We will call a real-valued field ϕ *simple* if there exist a finite sequence of non-negative numbers $0 \le t_1 < t_2 < \ldots < t_{k+1}$, a sequence ϕ_j of square integrable $\mathcal{F}_{t_j}$-measurable random variables and a sequence Γ_j of subsets of $\mathbb{R}^d$ such that $(t_j, t_{j+1}] \times \Gamma_j \in \mathcal{E}_0$. For simple fields ϕ, we define

$$\int_0^t \int_{\mathbb{R}^d} \phi(s, x)\, \check{\pi}(ds, dx) = \sum_{j=1}^k \phi_j\, \check{\pi}((t_j \wedge t, t_{j+1} \wedge t] \times \Gamma_j).$$

We have the following *isometric formula*:

PROPOSITION 8.4.2. *For simple random fields* ϕ:

$$(8.4.4) \qquad \mathbb{E}\left| \int_0^t \int_{\mathbb{R}^d} \phi(s, x)\, \check{\pi}(ds, dx) \right|^2 = \mathbb{E} \int_0^t \int_{\mathbb{R}^d} \phi^2(s, x)\, v(ds, dx).$$

PROOF. As in the proof of the isometric formula for the Wiener process we limit our calculations to $k = 1$. Let $0 \le s < t$, ϕ be $\mathcal{F}_s$-measurable random variable and $\mu((s, t] \times \Gamma) < +\infty$. Then,

$$\mathbb{E}\left(|\phi\, \check{\pi}((s, t] \times \Gamma)|^2\right) = \mathbb{E}\left(\phi^2\, \mathbb{E}\left(|\check{\pi}((s, t] \times \Gamma)|^2\right)\right) = (\mathbb{E}\phi^2)\, \mu((s, t] \times \Gamma),$$

so (8.4.4) holds in this case. $\qquad\qquad\qquad\qquad\qquad\qquad\qquad\qquad\qquad\square$

The isometric formula allows to extend the concept of integral to all those random fields which are limits, in the norm defined by the right hand side of (8.4.4), of simple random fields ϕ. The formula (8.4.4) remains true for them and the stochastic integral is again a square integrable martingale. Doob's regularisation theorem allows to claim that the stochastic integral

$$\int_0^t \int_{\mathbb{R}^d} \phi(s, x)\, \check{\pi}(ds, dx)$$

has a càdlàg version.

If ϕ is a simple field and

$$(8.4.5) \qquad \mathbb{E} \int_0^t \int_{\mathbb{R}^d} |\phi(s,x)| \, \mu(ds, dx) < +\infty, \qquad t \geq 0,$$

then the stochastic integral

$$(8.4.6) \qquad \int_0^t \int_{\mathbb{R}^d} \phi(s,x) \, \pi(ds, dx), \qquad t \geq 0,$$

can be defined, starting again from simple fields ϕ, but it is not a martingale. Thus if ϕ is a limit, in all the norms (8.4.5), of simple fields then the stochastic integral is a well defined process as the limit of stochastic integrals.

Note that processes $M(t) = \pi([0,t], \cdot)$ or $M(t) = \pi([0,t], \cdot) - t\mu(\cdot)$, are measure valued and have independent increments. Therefore, by choosing as U a proper Hilbert space of distributions on $\mathbb{R}^d$, the integrals (8.4.3), (8.4.6) can be treated as integrals

$$\int_0^t \psi(s) dM(s), \qquad t \geq 0$$

introduced in Chapter 9, where

$$\psi(s)\nu = \int_{\mathbb{R}^d} \phi(s,x)\nu(dx),$$

for $s \geq 0$ and any σ-finite measure ν on $\mathbb{R}^d$.

8.4.3. – Determining equations

To obtain Markov processes with general characteristics (F, Q, ν) one needs Ito equations of the form:
(8.4.7)
$$dX(t) = f(X(t)) \, dt + B(X(t)) \, dW(t)$$
$$+ \int_{\mathbb{R}^d} G_0(X(t-), y) \, \check{\pi}(dt, dy) + \int_{\mathbb{R}^d} G_1(X(t-), y) \, \pi(dt, dy),$$
$$X(0) = x$$

with properly chosen coefficients and random measure. The equation (8.4.7) should be understood as an integral equation
(8.4.8)
$$X(t) = x + \int_0^t f(X(s)) \, ds + \int_0^t B(X(s)) \, dW(s)$$
$$+ \int_0^t \int_{\mathbb{R}^d} G_0(X(s-), y) \, \check{\pi}(ds, dy) + \int_0^t \int_{\mathbb{R}^d} G_1(X(s-), y) \, \pi(ds, dy),$$

and a solution X should be a càdlàg, $(\mathcal{F}_t)$-adapted process.

Assume now that characteristics (f, Q, v) are given. How the coefficients of the equation (8.4.7) should be defined to obtain a solution X with the given characteristics? Following Ito's 1951 paper we write the operator A, from Theorem 8.2.1, in a more convenient form:

$$A\phi(x) = \langle f(x), D\phi(x) \rangle + \frac{1}{2} \text{ Trace } [Q(x)D^2\phi(x)]$$

$$+ \int_{|x|<1} (\phi(x+y) - \phi(x) - \langle y, D\phi(x) \rangle) \, v(x, dy)$$

$$+ \int_{|x|\geq 1} (\phi(x+y) - \phi(x)) \, v(x, dy),$$

where

$$f(x) = F(x) + \int_{|y|<1} \frac{y\,|y|^2}{1+|y|^2} \, v(x, dy) - \int_{|y|\geq 1} \frac{y}{1+|y|^2} \, v(x, dy).$$

Let π be the Poissonian measure corresponding to the intensity measure

$$\mu(ds, dx) = ds \, \frac{1}{|x|^{d+1}} \, dx.$$

Assume that for each x there exists a transformation $G(x, y)$ from $\mathbb{R}^d \backslash \{0\}$ into $\mathbb{R}^d$ such that

$$|G(x, y)| < 1 \text{ for } |y| < 1,$$
$$|G(x, y)| \geq 1 \text{ for } |y| \geq 1,$$

and that the image of the measure $|y|^{-d-1}dy$ by the transformation $G(x, \cdot)$ is precisely $v(x, \cdot)$. Define

$$G_0(x, y) = \begin{cases} G(x, y), & \text{if } |y| < 1, \\ 0 & \text{otherwise,} \end{cases}$$

$$G_1(x, y) = \begin{cases} G(x, y), & \text{if } |y| \geq 1, \\ 0 & \text{otherwise.} \end{cases}$$

Let in addition, for a given matrix valued function B, $B(x)RB^*(x) = Q(x)$, $x \in \mathbb{R}^d$. Then, under appropriate regularity conditions on f, B, G_0, G_1, the equation (8.4.7) has a unique Markovian solution with the characteristics (F, Q, v). We do not specify conditions under which the described representation holds but refer to Ito's papers and to the third volume of *The Theory of Stochastic Processes* by I. I. Gihman and A. V. Skorohod [29].

Let us finally notice that the Poissonian measure, used in the construction, is exactly the one which defines a Lévy process with the exponent function $\psi(\lambda) = |\lambda|$, $\lambda \in \mathbb{R}^d$. This process is called a *symmetric Cauchy process*.

Therefore, with some abuse of language, we can say that Markov processes can be effectively constructed from uniform Brownian and Cauchy motions.

REMARK. Treating processes π and $\check{\pi}$ as measure valued processes with independent increments, see the end of the previous subsection, the equation (8.4.7) can be rewritten as

$$dX(t) = H(X(t-))dZ(t), \quad X(0) = x$$

where the process Z has independent increments and is composed of 4 components:

$$t, \qquad W(t), \qquad \pi(t, \cdot), \qquad \check{\pi}(t, \cdot), \qquad t \geq 0.$$

Moreover

$$H(x)(\alpha, z, v_0, v_1) = \alpha f(x) + B(x)z + \int_{\mathbb{R}^d} G_0(x, z)v_0(dz)$$

$$+ \int_{\mathbb{R}^d} G_1(x, z)v_1(dz),$$

for real numbers and vectors α, z and measures v_0, v_1. This is in full analogy with the discrete time case, see equation (8.1.3).

CHAPTER 9

Stochastic integration in infinite dimensions

Of great interest in applications are dynamical systems for which the state space is infinite dimensional. Specific examples are provided by heat and reaction-diffusion processes or by electromagnetic waves. Deterministic dynamical systems of this type are described as solutions of partial differential equations of evolutionary types. They can be regarded as solutions of differential equations in infinite dimensional spaces. Stochastic dynamical systems in infinite dimensions, or equivalently, Markov processes in infinite dimensions, can be constructed using appropriate generalisation of the theory of stochastic integration in Hilbert spaces. To treat stochastic evolution equations with respect to both continuous and discontinuous noise processes it is convenient to develop the theory of stochastic integration with respect to square integrable Hilbert-valued martingales. To motivate better the material we start from more classical results when the martingale is a Wiener process.

9.1. – Integration with respect to a Wiener process

9.1.1. – Wiener process on a Hilbert space

Wiener processes which are used in applications evolve on a Hilbert or a Banach space U or also on linear topological spaces like the space of Schwartz distributions. The general definition is as follows. A stochastic process $W(t)$, $t \geq 0$, with values in E, is called a Wiener process if

1. $W(t)$, $t \geq 0$ has continuous paths and starts from 0.
2. The increments $W(t_2) - W(t_1), \ldots, W(t_n) - W(t_{n-1})$, for $0 \leq t_1 < t_2 < \ldots < t_n$ are independent random variables:

$$\mathbb{P}(W(t_2) - W(t_1) \in \Gamma_1, \dots, W(t_n) - W(t_{n-1}) \in \Gamma_{n-1})$$

(9.1.1)
$$= \prod_{i=1}^{n-1} \Pr\{W(t_{i+1}) - W(t_i) \in \Gamma_i\}.$$

3. For each measurable set $\Gamma \subset E$ and all $t, h \geq 0$

$$\mathbb{P}(W(t+h) - W(t) \in \Gamma) = \mathbb{P}(W(h) \in \Gamma).$$

In other words, an E-valued, continuous, stochastic process is a Wiener process if and only if for each x^* in the space E^* of all continuous linear functionals on E, the real valued process $\langle W(t), x^* \rangle$ is a 1-dimensional Wiener process. All processes and random variables are always assumed to be defined on a fixed probability space $(\Omega, \mathcal{F}, \Pr)$.

It is well known that 1-dimensional Wiener processes have Gaussian distributions. A probability measure μ on $\mathbb{R}^1$ is said to be Gaussian if it is either concentrated at a point $m \in \mathbb{R}^1$ or has a density

$$\frac{1}{\sqrt{2\pi q}} e^{-\frac{1}{2q}(x-m)^2}, \quad x \in \mathbb{R}^1.$$

It is denoted by $N(m, q)$. If $q = 0$, $N(m, 0) = \delta_{\{m\}}$. More generally a probability measure μ on a linear topological space is Gaussian if every functional $x^* \in E^*$, considered as a random variable on $(E, \mathcal{B}(E), \mu)$, is a real Gaussian random variable. Equivalently if linear functionals transport μ onto some $N(m, q)$.

We will additionally require that random variables $W(t)$ are symmetric that is,

$$\mathbb{E}\langle W(t), x^* \rangle = 0, \quad t \geq 0, \ x^* \in E^*.$$

By far the most important Wiener processes are those which evolve on Hilbert spaces. In a sense all other Wiener processes are variants of Hilbert space valued Wiener processes. Basic properties of Gaussian Hilbert space valued random variables are gathered in the following theorem.

THEOREM 9.1.1. *Assume that ξ is a symmetric Gaussian random variable with values in a separable Hilbert space H. Then for sufficiently small $s > 0$,*

$$\mathbb{E}(e^{s|\xi|^2}) < +\infty.$$

In particular all moments $\mathbb{E}|\xi|^k < +\infty$, $k = 1, 2, \dots$ are finite.

It follows from the theorem that $\mathbb{E}|\xi| \langle +\infty$ and $E\xi = 0$ and there exists a non-negative operator Q such that

$$\mathbb{E}\langle \xi, a \rangle \langle \xi, b \rangle = \langle Qa, b \rangle.$$

A non-negative operator Q is said to be trace class if and only if for an arbitrary orthonormal complete sequence (e_n):

$$\sum_{n=1}^{+\infty} \langle Qe_n, e_n \rangle < +\infty.$$

We have also

THEOREM 9.1.2. *The covariance operator Q is compact and trace class. For arbitrary $m \in H$ and arbitrary trace class non-negative operator Q there exists a unique Gaussian measure corresponding to (m, Q) and denoted by $N(m, Q)$ or $N_{m,Q}$.*

PROOFS.

LEMMA 9.1.3. *Assume that ξ takes values in $\mathbb{R}^d$, is Gaussian and such that $\mathbb{E}\xi = 0$. Then for all*

$$s < \frac{1}{2\mathbb{E}|\xi|^2}, \quad \text{we have} \quad \mathbb{E}(e^{s|\xi|^2}) \leq \frac{1}{\sqrt{1 - 2s\mathbb{E}|\xi|^2}}.$$

PROOF. We consider first the case $s < 0$ and denote $s = -\lambda$, $\eta = \sqrt{\lambda}\xi$. Then $\mathbb{E}e^{-\lambda|\xi|^2} = \mathbb{E}e^{-|\eta|^2}$. Let the matrix Q be the covariance of η, and $e_1, \ldots, e_d$ an orthonormal basis such that

$$Qe_j = \lambda_j e_j, \quad j = 1, \ldots, d$$

where $\lambda_1, \ldots, \lambda_d$ are the eigenvalues of Q. Then

$$\eta = \sum_{j=1}^{d} \langle \eta, e_j \rangle e_j = \sum_{j=1}^{d} \eta_j e_j$$

where $\eta_1, \ldots, \eta_d$ are independent with $N0, \lambda_1), \ldots, N(0, \lambda_d)$ distributions. Consequently

$$\mathbb{E}(e^{-|\eta|^2}) = \mathbb{E}(e^{-\sum_{j=1}^{d} \eta_j^2}) = \prod_{j=1}^{d} \mathbb{E}(e^{-\eta_j}) = \prod_{j=1}^{d} \frac{1}{\sqrt{1 + 2\lambda_j}}$$

$$= \frac{1}{\sqrt{(1 + 2\lambda_1) \cdots (1 + 2\lambda_d)}} \leq \frac{1}{\sqrt{1 + 2(\lambda_1 + \cdots \lambda_d)}}.$$

Since $\mathbb{E}|\eta|^2 = \lambda_1 + \cdots + \lambda_d$ the result follows. The proof for $s \in (0, \frac{1}{2\mathbb{E}|\xi|^2})$ is similar. $\qquad\square$

PROOF OF THEOREM 9.1.2. Choose an arbitrary orthonormal basis (e_n). Then

$$\xi = \sum_{n=1}^{+\infty} <\xi, e_n> e_n$$

and $\xi = \lim_d \xi_d$ where $\xi_d = \sum_{n=1}^{d} <\xi, e_n> e_n$. Moreover ξ_d is a d-dimensional Gaussian random variable and by the lemma

$$\mathbb{E}(e^{-|\xi_d|^2}) \leq \frac{1}{\sqrt{1 + 2\mathbb{E}|\xi_d|^2}}.$$

Note that $|\xi_d|^2 \uparrow |\xi|^2$ so $\mathbb{E}|\xi_d|^2 \uparrow \mathbb{E}|\xi|^2$ and therefore

$$\mathbb{E}(e^{-|\xi|^2}) \leq \frac{1}{\sqrt{1 + 2\mathbb{E}|\xi|^2}}.$$

If $\mathbb{E}|\xi|^2 = +\infty$ then $|\xi|^2 = +\infty$ with probability 1, which is not the case, so $\mathbb{E}|\xi|^2 < +\infty$. But then the lemma is true also for infinite dimensional Gaussian random variables using the fact that $|\xi_d|^2 \uparrow |\xi|^2$. This completes the proof of Theorem 9.1.2. $\qquad\square$

Since

$$\mathbb{E}|\xi|^2 = \mathbb{E}\left(\sum_n \langle \xi, e_n \rangle^2\right) = \sum_n \mathbb{E}\langle \xi, e_n \rangle^2 = \sum_n \langle Q e_n, e_n \rangle.$$

Trace $Q < +\infty$.

Thus distributions of any Hilbert space valued Wiener process W are uniquely determined by a non-negative trace class operator Q being the covariance of $W(1)$. Moreover

$$W(t) = \sum_n^{+\infty} \sqrt{\lambda_n} \beta_n(t) e_n \quad t \geq 0$$

where $\beta_1, \beta_2, \ldots$ are independent real Wiener processes and (λ_n, e_n) is the eigensequence of Q.

We assume now that $H = L^2(\mathcal{O}, \mu)$ where μ is a finite, non-negative measure. Then the operator Q is an integral operator

$$Q\varphi(x) = \int_{\mathcal{O}} q(x, y)\varphi(y)\mu(dy)$$

and the (positive definite) function q is the spatial correlation. If $W(t, x)$, $t \geq 0, x \in \mathcal{O}$ is the Q-Wiener process then

$$\mathbb{E}W(t, x)W(s, y) = (t \wedge s)q(x, y).$$

This heuristic formula can be justified in many specific cases.

EXAMPLES. Define

$$W(t, x) = \sum_{n}^{+\infty} \frac{\beta_n(t)}{n} \sqrt{\frac{2}{\pi}} \sin nx \quad x \in (0, \pi).$$

Then $q(x, y) = x \wedge y - \frac{1}{\pi}xy$, $x, y \in (0, \pi)$ and for each fixed t, $W(t, \cdot)$ is a Brownian bridge process.

Define

$$\widetilde{W}(t, x) = \sum_{n}^{+\infty} \frac{\beta_n(t)}{n + \frac{1}{2}} \sqrt{\frac{2}{\pi}} \sin\left[\left(n + \frac{1}{2}\right)x\right] \quad x \in (0, \pi).$$

Then $\widetilde{q}(x, y) = x \wedge y$. For each fixed t, $\widetilde{W}(t, \cdot)$ is a real valued Brownian motion. The process $\widetilde{W}$ is also called Brownian sheet on $[0, +\infty) \times [0, \pi]$.

9.1.2. – Stochastic integration with respect to a Wiener process

To deal with stochastic equations one needs the concept of stochastic integrals:

$$\int_0^t \Phi(s)dW(s)$$

where $\Phi(s, \omega)$ are operators from U into another Hilbert space H.

Let us assume that the process W is defined on a probability space $(\Omega, \mathcal{F}, \mathbb{P})$ and that $\mathcal{F}_t$ is an increasing family of σ fields contained in $\mathcal{F}$ and such that $W(t)$ is $\mathcal{F}_t$ measurable, for any $t \geq 0$, and that for any $t \geq s \geq 0$, $W(t) - W(s)$ is independent of $\mathcal{F}_s$. If this is the case one says that W is a *Wiener process with respect* to $\mathcal{F}_t$. The following proposition follows directly from the definitions.

PROPOSITION 9.1.4. *Let W be a Wiener process with respect to $\mathcal{F}_t$ with the covariance operator Q. Then for arbitrary vectors $a, b \in U$ and arbitrary $0 \leq s \leq t \leq u \leq v$*

$$(9.1.3) \quad \mathbb{E}(\langle W(t) - W(s), a\rangle\langle W(t) - W(s), b\rangle | \mathcal{F}_s) = (t - s)\langle Qa, b\rangle,$$

$$(9.1.4) \quad E(\langle W(t) - W(s), a\rangle\langle W(v) - W(u), b\rangle | \mathcal{F}_u) = 0.$$

Let $L(U, H)$ be the Banach space of linear continuous operators from U into H. An $L(U, H)$ stochastic process Φ is said to be simple if there exist a sequence of nonnegative numbers $t_0 = 0 < t_1 < \ldots < t_m$, a sequence of operators $\Phi_j \in L(U, H)$, $j = 1, \dots, m$, and a sequence of events $A_j \in \mathcal{F}_{t_j}$, $j = 0, \dots, m-1$, such that

$$\Phi(s) = \sum_{k=0}^{m-1} \mathbf{1}_{A_k} \mathbf{1}_{]t_k, t_{k+1}]}(s), \quad s \in [0, +\infty[,$$

where $\mathbf{1}_B$ denotes the indicator function of the set B. For simple processes Φ we set:

$$\int_0^t \Phi(s)dW(s) = \sum_{k=0}^{m-1} \Phi_k(W(t_{k+1} \wedge t) - W(t_k \wedge t)), \quad t \in]0, +\infty[.$$

Let $L_{HS}(U, H)$ be the space of all Hilbert-Schmidt operators, from U into H, equipped with the Hilbert-Schmidt norm $\|\cdot\|_{L_{HS}(U,H)}$.

PROPOSITION 9.1.5. *For arbitrary simple process* Φ:

$$\mathbb{E} \left| \int_0^t \Phi(s)dW(s) \right|^2 = \mathbb{E} \int_0^t \|\Phi(s)Q^{1/2}\|^2_{L_{HS}(U,H)} ds, \quad t \geq 0.$$

PROOF. By the very definition

$$\mathbb{E} \left| \int_0^t \Phi(s)dW(s) \right|^2 = \mathbb{E} \left| \sum_{k=0}^{m-1} \mathbf{1}_{A_k} \Phi_k(W(t_{k+1} \wedge t) - W(t_k \wedge t)) \right|^2$$

$$= \mathbb{E} \sum_{k,l=0}^{m-1} \mathbf{1}_{A_k} \mathbf{1}_{A_l} \langle \Phi_k(W(t_{k+1} \wedge t) - W(t_k \wedge t)), \Phi_l(W(t_{l+1} \wedge t) - W(t_l \wedge t)) \rangle.$$

By formulas (9.1.4),

(9.1.5) $\mathbb{E}(< W(t_{k+1} \wedge t) - W(t_k \wedge t), a >< W(t_{l+1} \wedge t) - W(t_l \wedge t), b > |\mathcal{F}_{t_l}) = 0$,

if $k \neq l$ and is equal

(9.1.6) $(t_{k+1} \wedge t - t_k \wedge t) < Qa, b >$,

if $k = l$.

Let e_j be an orthonormal basis in U and $k \leq l$. We have

$I_{klj} = \mathbb{E}(\langle \Phi_k(W(t_{k+1} \wedge t) - W(t_k \wedge t)), e_j \rangle \langle \Phi_l(W(t_{l+1} \wedge t) - W(t_l \wedge t)), e_j \rangle |\mathcal{F}_{t_l})$

$\quad = \mathbb{E}(\langle (W(t_{k+1} \wedge t) - W(t_k \wedge t)), \Phi_k^* e_j \rangle \langle (W(t_{l+1} \wedge t) - W(t_l \wedge t)), \Phi_l^* e_j \rangle |\mathcal{F}_{t_l})$.

By formulas (9.1.4) if $k < l$,

$$\mathbb{E} I_{klj} \mathbf{1}_{A_k} \mathbf{1}_{A_l} = 0$$

and if $k = l$,

$$\mathbb{E} I_{kkj} \mathbf{1}_{A_k} = \mathbb{E} \mathbf{1}_{A_k} (t_{k+1} \wedge t - t_k \wedge t) \langle Q\Phi_k^* e_j, \Phi_k^* e_j \rangle.$$

But

$$\sum_j \langle Q\Phi_k^* e_j, \Phi_k^* e_j \rangle = \sum_j |Q^{1/2}\Phi_k^* e_j|^2 = |Q^{1/2}\Phi_k^*|^2_{L_{HS}(U,H)},$$

and since

$$|Q^{1/2}\Phi_k^*|^2_{L_{HS}(U,H)} = |\Phi_k Q^{1/2}|^2_{L_{HS}(U,H)},$$

the required identity follows by elementary calculations. □

Note that the space
$$\widetilde{U} = Q^{1/2}(U)$$
equipped with the scalar product
$$\langle a, b \rangle \widetilde{U} = \langle Q^{-1/2}a, Q^{-1/2}b \rangle_U,$$
is a Hilbert space.

Let $L_{HS}(\widetilde{U}, H)$ be the space of all Hilbert- Schmidt operators equipped with the Hilbert- Schmidt norm $\| \cdot \|_{HS} = \| \cdot \|_{L_{HS}(\widetilde{U},H)}$. By Proposition 9.1.5, for simple Φ,

$$(9.1.7) \qquad \mathbb{E} \left| \int_0^t \Phi(s)dW(s) \right|^2 = \mathbb{E} \int_0^t \|\Phi(s)\|_{HS}^2 ds, \quad t \in [0, T].$$

The definition of the stochastic integral and the above formula 9.1.7 extend to all adapted, $L_{HS}(\widetilde{U}, H)$ valued processes for which the right hand side of (9.1.7) is finite. By the same localisation procedure as in finite dimensions the concept of the stochastic integral and all its basic properties can be extended to all adapted $L_{HS}(\widetilde{U}, H)$- processes $\Phi(s)$ for which

$$\mathbb{P} \left(\int_0^t \|\phi(s)\|_{HS}^2 ds < +\infty, \quad t \geq 0 \right) = 1.$$

9.2. – Stochastic integration with respect to martingales

9.2.1. – Introduction

If U is a separable Hilbert space and $W(t)$, $t \geq 0$, a U-valued Wiener process with covariance Q, then the isometric formula holds,

$$\mathbb{E} \left| \int_0^t \Phi(s) \, dW(s) \right|^2 = \mathbb{E} \int_0^t \|\Phi(s)Q^{1/2}\|_{HS}^2 \, ds.$$

To prove the isometric formula in the general case, when the integrator is an U-valued square integrable martingale M, one needs to introduce bracket processes
$$\langle M, M \rangle_t, \langle\langle M, M \rangle\rangle_t.$$

The former is a real process and the latter an operator valued. It turns out that there exist an operator valued stochastic process Q_s, $s \geq 0$ such that,

$$(9.2.1) \qquad \langle\langle M, M \rangle\rangle_t = \int_0^t Q_s d\langle M, M \rangle_s, \quad t \geq 0.$$

The values of the process Q_s, $s \geq 0$ are non-negative, trace class operators. In the case when M is the Wiener process W with the covariance Q, then
$$\langle W, W \rangle_t = t\,\mathrm{Trace}\,Q, \quad \langle\langle W, W \rangle\rangle_t = t\,Q/\mathrm{Trace}\,Q, \quad t \geq 0.$$

The isometric formula in the general case was discovered by Métivier and G. Pistone [38] and is the form

$$(9.2.2) \qquad \mathbb{E} \left| \int_0^t \Phi(s) \, dM(s) \right|^2 = \mathbb{E} \int_0^t \|\Phi(s)Q_s^{1/2}\|_{HS}^2 \, d\langle M, M \rangle_s.$$

9.2.2. – Doob-Meyer decomposition

Assume that U is a separable Hilbert space, with norm $|\cdot|$ and scalar product $<\cdot,\cdot>$, and $(\Omega, \mathcal{F}, \mathbb{P})$ a probability space equipped with an increasing family of σ-fields $\mathcal{F}_t \subset \mathcal{F}$, $t \in I$. The set I will be either a bounded interval $[0, T]$ or $[0, \infty)$. A U-valued family of random variables $X(t)$, $t \in I$ is called an integrable process if

$$(9.2.3) \qquad \mathbb{E}\,|X(t)| < +\infty, \qquad t \in I.$$

If for all $t \in I$, $X(t)$ is an $\mathcal{F}_t$-measurable random variable, then the family X is called an adapted process. An integrable and adapted U-valued process $X(t)$, $t \in I$, is said to be a martingale if

$$(9.2.4) \qquad \mathbb{E}\,(X(t)\,|\,\mathcal{F}_s) = X(s), \qquad \mathbb{P} - \text{a.s.}$$

for arbitrary $t, s \in I$, $t \geq s$, see Section 4.1. The identity (9.2.4) is equivalent to the statement

$$(9.2.5) \qquad \int_F X(t)\,d\mathbb{P} = \int_F X(s)\,d\mathbb{P}, \qquad F \in \mathcal{F}_s,\ s \leq t,\ s, t \in I.$$

A real valued integrable and adapted process $X(t)$, $t \in I$ is said to be a submartingale (resp. a supermartingale) if, see Section 4.1,

$$\mathbb{E}\,(X(t)\,|\,\mathcal{F}_s) \geq X(s), \qquad (\text{resp. } \mathbb{E}\,(X(t)\,|\,\mathcal{F}_s) \leq X(s)), \qquad \mathbb{P} - \text{a.s.}$$

LEMMA 9.2.1.

i) *If* $M(t)$, $t \in [0, T]$ *is a* U-valued martingale, then $|M(t)|$, $t \in [0, T]$ *is a submartingale.*

ii) *If* g *is an increasing, convex function from* $[0, +\infty)$ *into* $[0, +\infty)$ *and* $\mathbb{E}\,g(|M(t)|) < +\infty$ *for* $t \in [0, T]$, *then* $g(|M(t)|)$, $t \in [0, T]$ *is a submartingale.*

PROOF.

i) Let $t, s \in [0, T]$, $t > s$, then

$$|M(s)| = |\mathbb{E}\,(M(t)\,|\,\mathcal{F}_s)| \leq \mathbb{E}\,(|M(t)|\,|\,\mathcal{F}_s)$$

as required.

ii) Since $|M(s)| \leq \mathbb{E}\,(|M(t)|\,|\,\mathcal{F}_s)$, $\mathbb{P}$-a.s., $s < t$, then monotonicity and convexity of g together with Jensen's inequality imply:

$$(9.2.6) \qquad g(|M(s)|) \leq g(\mathbb{E}\,(|M(t)|\,|\,\mathcal{F}_s)) \leq \mathbb{E}\,(g(|M(t)|)\,|\,\mathcal{F}_s)$$

as required. $\square$

From now on we will assume that the martingale $M(t)$, $t \in [0, T]$ is right-continuous and square-integrable:

$$\mathbb{E} |M(t)|^2 < +\infty, \qquad t \in [0, T].$$

Then by the lemma, $|M(t)|^2$, $t \in [0, T]$ is a submartingale. Denote by $\mathcal{P}_T$ the smallest σ-field of subsets of $[0, T] \times \Omega$ containing all sets of the form: $(s, t] \times \Gamma$, where $0 \leq s < t \leq T$ and $\Gamma \in \mathcal{F}_s$, and also $\{0\} \times \Gamma$ where $\Gamma \in \mathcal{F}_0$. A stochastic process $X(t)$, $t \in [0, T]$ with values in a measurable space is called predictable if the function $X(t, \omega)$, $t \in [0, T]$, $\omega \in \Omega$ is measurable with respect to the σ-field $\mathcal{P}_T$. We will need the following fundamental result, see e.g. [53].

THEOREM 9.2.2 (Doob-Meyer decomposition). *For arbitrary càdlàg real valued submartingale Y there exists a unique predictable and right-continuous increasing process $A(t)$, $t \in [0, T]$ such that $A(0) = 0$ and*

(9.2.7) $$M(t) = Y(t) - A(t), \qquad t \in [0, T]$$

is a martingale.

We give only a proof of the discrete time version of the theorem.

PROPOSITION 9.2.3. *Assume that $Y_0, Y_1, \ldots, Y_N$ is a submartingale with respect to $\mathcal{F}_0, \mathcal{F}_1, \ldots, \mathcal{F}_N$. Then*

(9.2.8) $$Y_n = M_n + A_n, \qquad n = 0, 1, \ldots, N$$

where (M_n) is an $(\mathcal{F}_n)$ martingale and (A_n) is an increasing sequence such that

(9.2.9) $$A_0 = 0, \qquad A_n \text{ is } \mathcal{F}_{n-1} \text{ measurable for } n = 1, 2, \ldots, N.$$

The decomposition (9.2.8) with the specified properties is unique.

PROOF. Define

$$A_0 = 0 \qquad \text{and} \qquad A_n = \sum_{k=0}^{n} (\mathbb{E}(Y_k \mid \mathcal{F}_{k-1}) - Y_{k-1}), \qquad n = 1, 2, \ldots, N.$$

Since (Y_k) is a submartingale we have

$$\mathbb{E}(Y_k \mid \mathcal{F}_{k-1}) \geq Y_{k-1}, \qquad k = 1, 2, \ldots, N$$

and therefore the sequence (A_n) is increasing. It is also clear that A_n is $\mathcal{F}_{n-1}$ measurable. If now

$$M_n = Y_n - A_n, \qquad n = 0, 1, \ldots, N,$$

then, for $n = 1, 2, \ldots, N$,

$$M_n - M_{n-1} = Y_n - Y_{n-1} - (A_n - A_{n-1}) = Y_n - \mathbb{E}(Y_n \mid \mathcal{F}_{n-1}).$$

Consequently

$$\mathbb{E}(M_n - M_{n-1} \mid \mathcal{F}_{n-1}) = \mathbb{E}(Y_n \mid \mathcal{F}_{n-1}) - \mathbb{E}(Y_n \mid \mathcal{F}_{n-1}) = 0,$$

and therefore the sequence M_n, $n = 0, 1, \ldots, N$ is a martingale as required. To show uniqueness assume that

$$Y_n = M'_n + A'_n, \qquad n = 0, 1, \ldots, N,$$

for a martingale (M') and an increasing sequence (A'_n) satisfying (9.2.9). Then

$$(9.2.10) \qquad M_n - M'_n = A'_n - A_n, \qquad n = 0, 1, \ldots, N$$

and

$$M_{n-1} - M'_{n-1} = \mathbb{E}(M_n - M'_n \mid \mathcal{F}_{n-1}) = \mathbb{E}(A'_n - A_n \mid \mathcal{F}_{n-1}) = A'_n - A_n,$$

because $A'_n - A_n$ is $\mathcal{F}_{n-1}$ measurable. By (9.2.10) we therefore have

$$M_{n-1} - M'_{n-1} = M_n - M'_n, \qquad n = 1, 2, \ldots, N,$$

since $M_0 = M'_0 = Y_0$ the uniqueness follows. $\qquad\qquad\qquad\qquad\square$

If $M(t)$, $t \in [0, T]$ is a square-integrable and right-continuous martingale, with values in a separable Hilbert space U, then $|M(t)|^2$, $t \in [0, T]$ is a right-continuous real valued submartingale and the corresponding predictable process is denoted by $\langle M, M \rangle_t$, $t \in [0, T]$ and called the (angle) *bracket process*. If M and N are two square-integrable right-continuous martingales then there exists a unique right-continuous, predictable process denoted by $\langle M, N \rangle_t$, $t \in [0, T]$ with bounded variation, such that

$$<M(t), N(t)> -\langle M, N \rangle_t, \qquad t \in [0, T]$$

is a martingale. It is given by a polarisation identity,

$$\langle M, N \rangle_t = \frac{1}{2} \left(\langle M + N, M + N \rangle_t - \langle M, M \rangle_t - \langle N, N \rangle_t \right), \qquad t \in [0, T].$$

EXAMPLE 9.2.4. Assume that $M(t) = W(t)$, $t \geq 0$ is a U-valued Wiener process. Then

$$\langle W, W \rangle_t = t \operatorname{Tr} Q,$$

where Q is the covariance operator of $W(1)$.

In fact, for $0 \leq s \leq t \leq T$, by the independence of the increments,

$$\mathbb{E}(|W(t)|^2 \mid \mathcal{F}_s) = \mathbb{E}(|(W(t) - W(s)) + W(s)|^2 \mid \mathcal{F}_s)$$
$$= \mathbb{E}((|W(t) - W(s)|^2 + |W(s)|^2 + 2 <W(t) - W(s), W(s)>) \mid \mathcal{F}_s)$$
$$= \mathbb{E}|W(t) - W(s)|^2 + |W(s)|^2$$
$$= (t - s) \operatorname{Tr} Q + |W(s)|^2.$$

Consequently

$$\mathbb{E}(|W(t)|^2 - t \operatorname{Tr} Q \mid \mathcal{F}_s) = |W(s)|^2 - s \operatorname{Tr} Q,$$

as required.

With the same proof we have a more general result.

PROPOSITION 9.2.5. *Assume that $M(t)$, $t \in [0, T]$ is a square-integrable, right-continuous process, having independent, time-homogeneous increments, with respect to the family $(\mathcal{F}_t)$, $t \in [0, T]$, and starting from 0. Then*

$$\langle M, M \rangle_t = t \operatorname{Tr} Q,$$

where Q is a trace class, non-negative operator on U such that

(9.2.11) $\qquad <Qa, b> = \mathbb{E}[<M(1), a><M(1), b>], \qquad a, b \in U.$

Note that the relation (9.2.11) defines the operator Q uniquely. Its symmetricity and non-negativity is obvious. Since, for an arbitrary, complete orthonormal basis (e_n) in U,

$$\operatorname{Tr} Q = \sum_n <Qe_n, e_n> = \sum_n \mathbb{E}(<M(1), e_n>^2)$$

$$= \mathbb{E}\left(\sum_n <M(1), e_n>^2\right) = \mathbb{E}|M(1)|^2 < +\infty,$$

the trace of Q is finite. As for the Wiener process one shows that $\mathbb{E}|M(t)|^2 = t\mathbb{E}|M(1)|^2$ for $t \geq 0$.

9.2.3. – Operator valued angle bracket process

For stochastic integration we need a generalisation of the angle bracket. Denote by $L_1(U)$ the space of all nuclear operators on U equipped with the standard nuclear norm, see e.g. [13]. Then $L_1(U)$ is a separable Banach space. By $a \otimes b$ we denote a rank one operator on U given by:

$$a \otimes b\,(u) = a <b, u>, \qquad u \in U.$$

It is easy to show that $\|a \otimes b\|_{L_1(U)} = |a|\,|b|$, $a, b \in U$. By $L_1^+(U)$ we denote the subspace of $L_1(U)$ consisting of all self-adjoint, non-negative, nuclear operators. If $M(t)$, $t \in [0, T]$ is a right-continuous, square-integrable process then the process

$$M(t) \otimes M(t), \qquad t \in [0, T],$$

is an $L_1(U)$ valued, right-continuous process such that

$$\mathbb{E}\,\|M(t) \otimes M(t)\|_{L_1(U)} = \mathbb{E}|M(t)|^2 \leq \mathbb{E}|M(T)|^2 < +\infty, \qquad t \in [0, T].$$

We have the following basic result, mentioned in the introduction to the section, from which the isometric formula will follow.

THEOREM 9.2.6. *There exists a unique right-continuous, $L_1^+(U)$ valued, increasing, predictable process $\langle\langle M, M\rangle\rangle_t$, $t \in [0, T]$, $\langle\langle M, M\rangle\rangle_0 = 0$ such that the process*

$$M(t) \otimes M(t) - \langle\langle M, M\rangle\rangle_t, \qquad t \in [0, T],$$

is an $L_1(U)$ martingale. Moreover there exists a predictable $L_1^+(U)$ valued process Q_t, $t \in [0, T]$ such that

$$(9.2.12) \qquad\qquad \langle\langle M, M\rangle\rangle_t = \int_0^t Q_s \, d\langle M, M\rangle_s.$$

PROOF. Let (e_k) be an orthonormal and complete basis in U. Then

$$M(t) = \sum_{k=1}^{+\infty} <M(t), e_k> e_k, \qquad t \in [0, T].$$

Denote the process $<M(t), e_k>$, $t \in [0, T]$ by $M^k(t)$, $t \in [0, T]$. Since

$$\mathbb{E} |M(t)|^2 = \mathbb{E} \sum_{k=1}^{+\infty} (M^k(t))^2 = \sum_{k=1}^{+\infty} \mathbb{E} (M^k(t))^2 < +\infty, \qquad t \in [0, T],$$

the processes M^k are right-continuous, square-integrable, real martingales. Let $\langle M^k, M^j\rangle$, $k, j = 1, 2, \ldots$ be the corresponding angle bracket processes. Then one should have

$$(9.2.13) \qquad\qquad \langle\langle M, M\rangle\rangle_t = \sum_{k,l=1}^{+\infty} e_k \otimes e_l \, \langle M^k, M^l\rangle_t, \qquad t \in [0, T].$$

To see that (9.2.13) defines an $L_1^+(U)$ valued process we show that the series is a Hilbert-Schmidt operator.

In fact

$$\left\| \sum_{k,l} e_k \otimes e_l \langle M^k, M^l\rangle_t \right\|_{HS}^2 = \sum_{k,l} \left| \langle M^k, M^l\rangle_t \right|^2$$

$$\leq \sum_{k,l} \langle M^k, M^k\rangle_t \langle M^l, M^l\rangle_t$$

$$\leq \left(\sum_k \langle M^k, M^k\rangle_t \right)^2 < +\infty,$$

because $\mathbb{E} \sum_k \langle M^k, M^k\rangle_t = \mathbb{E} \sum_k (M^k(t))^2 = \mathbb{E}|M(t)|^2 < +\infty$.

The inequality $|\langle M^k, M^k\rangle_t|^2 \leq \langle M^k, M^k\rangle_t \langle M^k, M^k\rangle_t$, needed for the estimate, follows from the proof of the lemma below. This proves that the operator-valued function $\langle\langle M, M\rangle\rangle$ is a well defined $L_2(U)$ valued function. The operators $\langle\langle M, M\rangle\rangle_t$, $t \in [0, T]$ are symmetric and non-negative because

$$\langle <a, M>, <b, M> \rangle_t = \langle <b, M>, <a, M> \rangle_t$$

and

$$\langle <a, M>, <a, M> \rangle_t \geq 0, \qquad t \in [0, T].$$

For $0 \leq s \leq t \leq T$ we have

$$<\{\langle\langle M, M\rangle\rangle_t - \langle\langle M, M\rangle\rangle_s\}a, a> = \langle <a, M>, <a, M> \rangle_t - \langle <a, M>, <a, M> \rangle_s \geq 0,$$

so the operators $\langle\langle M, M\rangle\rangle_t - \langle\langle M, M\rangle\rangle_s$ are non-negative. Consequently

$$\|\langle\langle M, M\rangle\rangle_t - \langle\langle M, M\rangle\rangle_s\|_{L_1(U)} = \mathrm{Tr}\,\{\langle\langle M, M\rangle\rangle_t - \langle\langle M, M\rangle\rangle_s\}$$
$$= \sum_j \{\langle M^j, M^j\rangle_t - \langle M^j, M^j\rangle_s\},$$

$$\mathbb{E}\,\|\langle\langle M, M\rangle\rangle_t - \langle\langle M, M\rangle\rangle_s\|_{L_1(U)} = \mathbb{E}\,(|M(t)|^2 - |M(s)|^2) < +\infty.$$

This shows that $\langle\langle M, M\rangle\rangle_t$, $t \in [0, T]$ is an $L_1^+(U)$ valued, predictable, increasing function. To show that it is also right-continuous and can be represented in the form (9.2.12) we establish first a lemma.

LEMMA 9.2.7. *Assume that M, N, are right-continuous, square-integrable martingales on $[0, T]$. There exists a predictable process $q(s)$, $s \in [0, T]$ such that*

$$\langle M, N\rangle_t = \int_0^t q(s)\, d[\langle M, M\rangle_s + \langle N, N\rangle_s].$$

PROOF. It is enough to show that for almost all random outcome the measure in $[0, T]$ corresponding to the function $\langle M, N\rangle$ of bounded variation is absolutely continuous with respect to the sum of the measures induced by $\langle M, M\rangle$ and $\langle N, N\rangle$. For fixed $s \in [0, T]$ and arbitrary real x, the process

$$\langle M + xN, M + xN\rangle_t - \langle M + xN, M + xN\rangle_s, \qquad t \in [s, T]$$

is the angle bracket corresponding to $M(t) + xN(t)$, $t \in [s, T]$. Consequently, for all $x \in \mathbb{R}^1$

$$\langle M + xN, M + xN\rangle_t - \langle M + xN, M + xN\rangle_s$$
$$= x^2(\langle N, N\rangle_t - \langle N, N\rangle_s) + 2x(\langle M, N\rangle_t - \langle M, N\rangle_s) + (\langle M, M\rangle_t - \langle M, M\rangle_s) \geq 0$$

and thus

$$(\langle M, N\rangle_t - \langle M, N\rangle_s)^2 \leq (\langle M, M\rangle_t - \langle M, M\rangle_s)(\langle N, N\rangle_t - \langle N, N\rangle_s)$$

or equivalently

(9.2.14) $|\langle M, N\rangle_t - \langle M, N\rangle_s| \le (\langle M, M\rangle_t - \langle M, M\rangle_s)^{1/2}(\langle N, N\rangle_t - \langle N, N\rangle_s)^{1/2}.$

From (9.2.14),

$$|\langle M, N\rangle_t - \langle M, N\rangle_s| \le \frac{1}{2}\{(\langle M, M\rangle_t + \langle N, N\rangle_t) - (\langle M, M\rangle_s + \langle N, N\rangle_s)\}.$$

This way we have shown that on each subinterval of $[0, T]$ the total variation of the measure corresponding to $\langle M, N\rangle$ is smaller than the total variation corresponding to $\langle M, M\rangle + \langle N, N\rangle$. In particular if a Borel set $\Gamma \subset [0, T]$ is of measure zero with respect to $d(\langle M, M\rangle + \langle N, N\rangle)$ it is of measure zero with respect to $d\langle M, N\rangle$ and this implies the required absolute continuity.

To prove predictability of q we use a real analysis result. If μ and ν are two finite non-negative measures on $[0, +\infty)$ and μ is absolutely continuous with respect to ν then for ν-almost all $t > 0$

(9.2.15) $$\frac{d\mu}{d\nu}(t) = \liminf_{r \uparrow t} \frac{\mu((r, t])}{\nu((r, t])},$$

where the limit is taken with respect to $r < t$, r rational. Since the limes inferior of predictable processes is predictable the result follows. □

It follows from the lemma that for each $i, j \in \mathbb{N}$ there exists a predictable process $q^{i,j}(t)$, $t \in [0, T]$ such that

$$\langle M^i, M^j\rangle_t = \int_0^t q^{i,j}(s)\, d\langle M, M\rangle_s = \int_0^t q^{i,j}(s)\, d\sum_{k=1}^{+\infty}\langle M^k, M^k\rangle_s.$$

Thus, if the space U is finite dimensional we have the representation

$$\langle\langle M, M\rangle\rangle_t = \int_0^t Q_s\, d\langle M, M\rangle_s$$

where Q_s, $s \in [0, T]$ is by (9.2.15) a predictable function with values in the space $L_1^+(U)$. In general we have for arbitrary $a \in U$ with only a finite number of coordinates different from zero

$$<\langle\langle M, M\rangle\rangle_t a, a> = \langle <a, M>, <a, M>\rangle_t = \int_0^t <Q_s a, a>\, d\langle M, M\rangle_s.$$

In particular

$$\text{Trace } \langle\langle M, M\rangle\rangle_t = \int_0^t \text{Trace } Q_s\, d\langle M, M\rangle_s < +\infty$$

$\mathbb{P}$-a.s., where Q_s is given by the formula $Q_s = \sum_{k,l=1}^{+\infty} e_k \otimes e_l\, q^{k,l}(s)$. This implies that Q_s takes values in $L_1^+(U)$ even if $\dim U = +\infty$. □

9.2.4. – Final comments

The proof of the isometric formula (9.2.2) is now similar to its proof for the Wiener process. The main ingredient is the following generalisation of Proposition 9.1.4 which follows directly from Theorem 9.2.6,

PROPOSITION 9.2.8. *Let M be a Wiener process with respect to $\mathcal{F}_t$ with the covariance operator Q. Then for arbitrary vectors $a, b \in U$ and arbitrary $0 \leq s \leq t \leq u \leq v$*

$$(9.2.16) \quad \mathbb{E}\langle <M(t)-M(s), a><M(t)-M(s)>, b\rangle|\mathcal{F}_s) = \int_s^t <Q_\sigma a, b> \, d\langle M, M\rangle_\sigma$$

$$(9.2.17) \quad E(<M(t) - M(s), a><M(u) - M(v), b> |\mathcal{F}_u) = 0.$$

As an introduction to *stochastic evolution equations* with respect to Hilbert valued martingales we comment on the so called *weak solution*.

Let H and U be separable Hilbert spaces, A the infinitesimal generator of a C_0-semigroup $S(t)$, $t \geq 0$ on H and $M(t)$, $t \geq 0$ a square integrable martingale on U. Consider an equation

$$(9.2.18) \quad dy(t) = Ay(t)dt + f(t)dt + \phi(t)dM(t), \quad t \geq 0, \ y(0) = x,$$

where f is an H-valued, adapted process and $\phi(t)$, $t \geq 0$ an adapted process of, possibly unbounded, linear operators from U into H. Both processes f, ϕ can be functionals of the unknown process y.

A linear subspace $D \subset H$ is said to be a core for $S(t)$, $t \geq 0$ if $D \subset D(A)$ and for every $x \in D(A)$ there exists a sequence (x_n) of elements from D such that $x_n \to x$ and $Ax_n \to Ax$ as $n \to +\infty$. It is well known, see [2], that if a dense linear subspace $D \subset H$ is invariant for the semigroup then it is a core. Let D^* be a core of the operator A^* adjoint to A. A continuous, adapted process $y(t)$, $t \geq 0$ is said to be *a weak solution* of (9.2.18) if for each $x^* \in D^*$ and $t \geq 0$, P-a.s.,

$$(9.2.19) \quad <x^*, y(t)> = <x^*, x> \int_0^t <A^*x^*, y(s)> \, ds$$

$$(9.2.20) \quad + \int_0^t <x^*, f(s)> \, ds + \int_0^t <\phi^*(s)x^*, dM(s)>_U \ .$$

We have the following important proposition which allows to replace differential equations by integral equations.

PROPOSITION 9.2.9. *Process*

$$(9.2.21) \quad y(t) = S(t)x + \int_0^t S(t-s)f(s)ds + \int_0^t S(t-s)\phi(s)dM(s)$$

is a weak solution of the equation (9.2.18).

The relation (9.2.21) is the so called *variation of constant formula*.

CHAPTER 10

Prohorov's theorem

The chapter is devoted to an exposition of Prohorov's tightness theorem in metric spaces.

10.1. – Motivations

The Kolmogorov theorem is of great theoretical importance. In practice however only seldom finite dimensional distributions are known. Two stochastic processes X and X' are called *equivalent* if they have identical finite dimensional distributions. For each $\omega \in \Omega$, the function $t \to X(t, \omega)$ is called *a path* of the process X. It is easy to construct two equivalent processes such that one of them has continuous paths and the paths of the other one are discontinuous.

EXAMPLE 10.1.1 Define $X(t) = t$, for $t \in [0, 1]$. Let $X'(t)$, $t \in [0, 1]$, be defined on the probability space $((0, 1], \mathcal{B}((0, 1]), \mathbb{P})$, where $\mathbb{P}$ is the Lebesgue measure on $(0, 1]$, by the formula: $X(t, \omega) = 0$ if $t = \omega$ and $X'(t, \omega) = t$ if $t \neq \omega$. Then the processes X and X' are equivalent although the trajectories of X are continuous and all the trajectories of X' are discontinuous. □

A priori any function from $\mathcal{T}$ into E can be a path of a process constructed in Kolmogorov's way. To prove that there exists an equivalent version of the process with special paths properties requires an additional work.

A powerful way of constructing good versions of stochastic processes is based on the concept of weak convergence of measures.

EXAMPLE 10.1.2 Let $\xi_1, \xi_2, \ldots$ be a sequence of independent, real, random variables with identical Gaussian laws with mean (expectation) 0 and second moment 1. Such a sequence exists on a probability space $(\Omega, \mathcal{F}, \mathbb{P})$. For each

$n = 1, 2, \ldots$ and $t \in [0, 1]$ define

$$S_n = \xi_1 + \ldots + \xi_n, \qquad S_0 = 0,$$

$$X_n(t) = \frac{1}{\sqrt{n}} S_{[nt]} + (nt - [nt]) \frac{1}{\sqrt{n}} \xi_{[nt]+1}.$$

For each $n = 1, 2, \ldots$ and $\omega \in \Omega$, $X_n(\cdot, \omega)$ is a continuous function and X_n can be regarded as a random variable with values in $C[0, 1]$. Let μ_n be the law of X_n treated as $C[0, 1]$-valued random variable. One can show that the sequence $\mu_n = \mathcal{L}(X_n)$ weakly converges to a measure μ on $C[0, 1]$ which can be identified with the so called Wiener measure. The measure μ is automatically concentrated on $C[0, 1]$. If one takes now as a new probability space $(C[0, 1], \mathcal{B}(C[0, 1]), \mu)$ and defines $X(t, \omega) = \omega(t)$ for $\omega \in C[0, 1]$, then X is the so called Wiener process with continuous paths.

Let (X_n) be a sequence of E-valued random variables. One says that (X_n) converges weakly to X if the corresponding distributions $\mathcal{L}(X_n) = \mu_n$ converge weakly to the law $\mathcal{L}(X) = \mu$. If $\mu_n \Rightarrow \mu$ and ϕ is a continuous function $\phi : E \to \mathbb{R}^1$ then also $(\phi(X_n))$ converges weakly to $\phi(X)$.

PROOF. For arbitrary $\lambda \in \mathbb{R}^1$,

$$\mathbb{E}(e^{i\lambda\phi(X_n)}) = \int_E e^{i\lambda\phi(x)} \mu_n(dx) \to \int_E e^{i\lambda\phi(x)} \mu(dx),$$

because $x \to e^{i\lambda\phi(x)}$ is a bounded continuous function. This implies that characteristic functions of the laws of real valued random variables $\phi(X_n)$ converge to the characteristic function of the law of $\phi(X)$. This proves the result. □

In particular if we know that continuous processes $X_n(t)$, $t \in [0, T]$, converge weakly in $C([0, T]; E)$ to a continuous process $X(t)$, $t \in [0, T]$, and ϕ is a continuous function on $C([0, T]; E)$ then $\phi(X_n)$ converges weakly to $\phi(X)$. Interesting functions are for instance the following ($E = \mathbb{R}^1$):

$$\phi(X) = \max_{0 \leq t \leq T} X(t), \qquad \phi(X) = \int_0^T X(t)\, dt.$$

10.2. – Weak topology

E is Polish, that is separable, complete, metric space, with metric ρ, $\mathcal{E}$ is the Borel σ-field of subsets of E and $\mathcal{M}_1(E)$ is the set of all probability measures on E. Denote by $C_b(E)$ the space of all bounded continuous functions on E. Then $\mathcal{M}_1(E)$ can be regarded as a subset of $C_b(E)^*$, the space of all bounded linear functionals on $C_b(E)$, equipped with the weakest topology such that all mappings $\phi \to \phi(f)$ are continuous for all $f \in C_b(E)$. We say that a sequence (μ_n) converges weakly to μ if an arbitrary $C_b(E)^*$ neighbourhood of

μ contains all but a finite number of elements of (μ_n). This can be equivalently stated as follows: for arbitrary $f \in C_b(E)$

(10.2.1) $(\mu_n, f) \to (\mu, f)$ as $n \to +\infty.$

PROPOSITION 10.2.1. *Let (μ_n) and μ be elements of $\mathcal{M}_1(E)$. The following conditions are equivalent.*

(i) $(\mu_n, f) \to (\mu, f)$ *for all* $f \in C_b(E)$.
(ii) $(\mu_n, f) \to (\mu, f)$ *for all* $f \in UC_b(E)$.
(iii) $\overline{\lim}_{n\to\infty}\mu_n(F) \le \mu(F)$ *for all closed sets* $F \subset E$.
(iv) $\underline{\lim}_{n\to\infty}\mu_n(G) \ge \mu(G)$ *for any open set* $G \subset E$.
(v) $\lim_{n\to\infty}\mu_n(A) = \mu(A)$ *for any* $A \in \mathcal{E}$ *such that* $\mu(\partial A) = 0$.

PROOF. $(i) \Rightarrow (ii)$ is obvious.

$(ii) \Rightarrow (iii)$. Let F be a given closed set and F_ϵ its ϵ-neighbourhood. Then there exists a uniformly continuous function f_ϵ on E such that $0 \le f_\epsilon \le 1$, $f_\epsilon = 1$ on F and $f_\epsilon = 0$ on F_ϵ^c. Define $f_\epsilon(x) = \phi(\frac{1}{\epsilon}\rho(x, F))$, where $\phi : \mathbb{R}^1 \to \mathbb{R}^1$ is piecewise linear, $\phi(\xi) = 1$ for $\xi \le 0$, $\phi(\xi) = 0$ for $\xi \ge 1$; also recall that $\rho(x, F) - \rho(y, F) \le \rho(x, y)$, $x, y \in E$. Then

$$\mu_n(F) = \int_F f_\epsilon \mu_n \le \int_E f_\epsilon \mu_n \to \int f_\epsilon \mu,$$

so $\overline{\lim}_{n\to\infty}\mu_n(F) \le \int_E f_\epsilon(x)\mu(dx)$, but $f_\epsilon \to \chi_F$ as $\epsilon \to 0$, so $\int_E f_\epsilon\mu \to \mu(F)$.

$(iii) \Leftrightarrow (iv)$. By taking into account that G^c is a closed set and $\nu(G^c) = 1 - \nu(G)$ for arbitrary $\nu \in \mathcal{M}_1(E)$.

$(iii) \Rightarrow (i)$. We will show that $\overline{\lim}_{n\to\infty}(\mu_n, f) \le (\mu, f)$. We can assume that $0 < f < 1$ (adding a constant and multiplying by a positive number). Fix $k \in \mathbb{N}$ and define closed sets

$$F_i = \left\{ x : f(x) \ge \frac{i}{k} \right\}, \qquad i = 0, 1, \dots, k.$$

Then F_i are closed sets and for arbitrary $\mu \in \mathcal{M}_1(E)$

$$\sum_{i=1}^k \frac{i-1}{k}\mu\left(x; \frac{i-1}{k} \le f < \frac{i}{k}\right) \le \int f\mu \le \sum_{i=1}^k \frac{i}{k}\mu\left(x; \frac{i-1}{k} \le f < \frac{i}{k}\right),$$

or equivalently

$$\sum_{i=1}^k \frac{i-1}{k}[\mu(F_{i-1}) - \mu(F_i)] \le \int f\mu \le \sum_{i=1}^k \frac{i}{k}[\mu(F_{i-1}) - \mu(F_i)].$$

Taking into account cancellation:

$$\frac{1}{k}\sum_{i=1}^k \mu(F_i) \le \int f\mu \le \frac{1}{k} + \frac{1}{k}\sum_{i=1}^k \mu(F_i).$$

Consequently

$$\overline{\lim}_{n\to\infty} \int f\,\mu_n \leq \frac{1}{k} + \frac{1}{k}\sum_{i=1}^{k}\mu(F_i) \leq \frac{1}{k} + \int f\,\mu.$$

Since $\overline{\lim}_{n\to\infty}(\mu_n, -f) \leq (\mu, -f)$ one gets also that $\underline{\lim}_{n\to\infty}(\mu_n, f) \geq (\mu, f)$. This proves (i).

Equivalence with (v). Recall that ∂A consists of all those points in E which have in any neighbourhood points in A and points in A^c. Then $\partial A = \overline{A} \cap \overline{A^c}$. Note also that $\overline{A} = \overset{\circ}{A} \cup \partial A$ and that $\partial A = \overline{A}\backslash\overset{\circ}{A}$. Assume that $\mu(\partial A) = 0$. Then

$$\mu(\overline{A}) \geq \overline{\lim_{n\to\infty}}\,\mu_n(\overline{A}) \geq \overline{\lim_{n\to\infty}}\,\mu_n(A) \geq \underline{\lim_{n\to\infty}}\,\mu_n(A) \geq \underline{\lim_{n\to\infty}}\,\mu_n(\overset{\circ}{A}) \geq \mu(\overset{\circ}{A}).$$

But if $\mu(\partial A) = 0$ then $\mu(\overline{A}) = \mu(\overset{\circ}{A})$ and from the above inequalities

$$\overline{\lim_{n\to\infty}}\,\mu_n(A) = \underline{\lim_{n\to\infty}}\,\mu_n(A).$$

Assume that for all Borel sets A such that $\mu(\partial A) = 0$ $\lim_{n\to\infty}\mu_n(A) = \mu(A)$. Now for arbitrary closed set F and its closed δ-neighbourhood $F_\delta = \{x : \rho(x, F) \leq \delta\}$ one has that $\partial F_\delta \subset \{x : \rho(x, F) = \delta\}$. For different $\delta > 0$ the sets $\{x : \rho(x, F) = \delta\}$ are disjoint and there exists a sequence $\delta_k \downarrow 0$ such that $\mu(x : \rho(x, F) = \delta_k) = 0$. Consequently $\lim_{n\to\infty}\mu_n(F_{\delta_k}) = \mu(F_{\delta_k})$. However $\mu_n(F) \leq \mu_n(F_{\delta_k})$ and therefore $\overline{\lim}_{n\to\infty}\mu_n(F) \leq \mu(F_{\delta_k})$ for each k. But $\mu(F_{\delta_k}) \downarrow \mu(F)$ so $\overline{\lim}_{n\to\infty}\mu_n(F) \leq \mu(F)$. This proves the result.   □

10.3. – Metrics on $\mathcal{M}_1(E)$

A metric space E equipped with a metric d will be denoted by E_d.

LEMMA 10.3.1. *Let E_ρ be a separable metric space. Then there exists a metric d equivalent to ρ such that the space $UC_b(E_d)$ is separable.*

PROOF. Let (a_k) be a dense countable subset of E. Define a mapping $h : E \to [0, 1]^{\mathbb{N}}$ by the formula:

$$h(x) = \left(\frac{\rho(x, a_k)}{1 + \rho(x, a_k)}\right).$$

Then h is continuous, one to one and the inverse is continuous. Thus E is homeomorphic to a subset $h(E)$ of the compact metric space $[0, 1]^{\mathbb{N}}$. Identifying E with $h(E)$, a metric that E inherits from $[0, 1]^{\mathbb{N}}$ is

$$d(x, y) = \sum_{k=1}^{+\infty} \frac{1}{2^k}\frac{\rho(x, a_k)}{1 + \rho(x, a_k)},$$

which is therefore equivalent to ρ.

Now let $\widehat{E}_d$ be the completion of E_d with respect to d. Then $\widehat{E}_d$ is compact and $UC_b(\widehat{E}_d) = C(\widehat{E}_d)$. But $C(\widehat{E}_d)$ is separable, by Lemma 10.3.2 below, and consequently $UC_b(\widehat{E}_d)$ is separable as well. Since $UC_b(\widehat{E}_d)$ is isomorphic to $UC_b(E_d)$ the result follows.   □

LEMMA 10.3.2. *If F is a compact metric space then the space $C(F)$ is separable.*

PROOF. We first note that the space $C([0,1]^{\mathbb{N}})$ is separable because, for arbitrary natural number k, $C([0,1]^k)$ is separable, and continuous functions on $[0,1]^{\mathbb{N}}$, which depend on a finite number of coordinates, are dense in $C([0,1]^{\mathbb{N}})$ by the Stone-Weierstrass theorem.

Next we construct a homeomorphism h of F onto a subset of $[0,1]^{\mathbb{N}}$, proceeding as in the proof of Lemma 10.3.1. Then $C(F)$ and $C(h(F))$ are isometric. Since F is compact, $h(F)$ is a compact subset of $[0,1]^{\mathbb{N}}$. Since every continuous function on $h(F)$ extends to a continuous function on $[0,1]^{\mathbb{N}}$, it is easily seen that separability of $C([0,1]^{\mathbb{N}})$ implies that $C(h(F))$ is separable as well, and the result follows.   □

PROPOSITION 10.3.3. *Let E be a Polish space. Then there exists a countable, dense subset of $C_b(E)$, denoted by (f_k), such that the following metric*

$$\Delta(\mu, \nu) = \sum_{k=1}^{+\infty} \frac{1}{2^k} \frac{|(\mu, f_k) - (\nu, f_k)|}{1 + |(\mu, f_k) - (\nu, f_k)|}$$

defines the weak topology on $\mathcal{M}_1(E)$.

PROOF. Let d be the metric constructed in Lemma 10.3.1. We take as (f_k) a countable, dense subset of $UC_b(E_d)$.

Let $\mathcal{O}$ be a $C_b(E)^*$ neighbourhood of μ. One has to show that there exists $r > 0$ such that $\{\nu : \Delta(\mu, \nu) < r\} \subset \mathcal{O}$. One can assume that $\mathcal{O}$ is determined by a finite number of continuous functions $g_1, \ldots, g_k \in C_b(E)$ and numbers $r_1, \ldots, r_k$:

$$\mathcal{O} = \{\nu : |(g_i, \mu) - (g_i, \nu)| < r_i, \ i = 1, 2, \ldots, k\}.$$

Assume to the contrary that there exists a sequence of measures ν_n such that $\Delta(\mu, \nu_n) \to 0$ but nevertheless $\nu_n \notin \mathcal{O}$. Without any loss of generality one can assume that

$$|(g_1, \mu) - (g_1, \nu_n)| \geq r_1$$

for all $n = 1, 2, \ldots$. But $(f_k, \nu_n) \underset{n \to \infty}{\to} (f_k, \mu)$ for a dense set in $UC_b(E_d)$ and therefore the convergence takes place for all f in $UC_b(E_d)$ and therefore for all f in $C_b(E)$, in particular for $f = g_1$, a contradiction. Fix now $r > 0$ and consider a ball of radius r with centre μ: $B(\mu, r) = \{\nu : \Delta(\mu, \nu) < r\}$. Take

$$\mathcal{O} = \{\nu : |(f_k, \mu) - (f_k, \nu)| < r_k, \ k = 1, 2, \ldots, N\}$$

and N such that $\sum_{k>N} 2^{-k} < r/2$. If $r_1, \ldots, r_N$ are chosen such that

$$\sum_{k=1}^{N} \frac{1}{2^k} \frac{r_k}{1 + r_k} < \frac{r}{2}$$

then $\mathcal{O} \subset B(\mu, r)$. This completes the proof. $\square$

REMARK 10.3.4. One can introduce an equivalent metric which is even complete. An example of such a metric was considered by Prohorov:

$$L(\mu, \nu) = \inf\{\delta : \mu(F) \le \nu(F_\delta) + \delta \text{ and } \nu(F) \le \mu(F_\delta) + \delta \text{ for all closed } F \subset E\}.$$

Another possibility is to use the so called Wasserstein metric, called also Fortet-Mourier metric

$$W(\mu, \nu) = \sup \left\{ \left| \int_E \phi(x)\, \mu(dx) - \int_E \phi(x)\, \nu(dx) \right| : \right.$$
$$\left. \sup_x |\phi(x)| \le 1, \ |\phi(x) - \phi(y)| \le \rho(x, y), \ x, y \in E \right\}.$$

10.4. – Prohorov's theorem

THEOREM 10.4.1.

(i) *Let* $\Gamma \subset \mathcal{M}_1(E)$ *be a compact set then for arbitrary* $\epsilon > 0$ *there exists a compact set* $K \subset E$ *such that*

(10.4.1) $\mu(K) \ge 1 - \epsilon \quad \text{for all } \mu \in \Gamma.$

(ii) *If* $\Gamma \subset \mathcal{M}_1(E)$ *is a set such that for each* $\epsilon > 0$ *there exists a compact set* $K \subset E$ *such that* (10.4.1) *holds then* $\overline{\Gamma}$ *is a compact subset of* $\mathcal{M}_1(E)$.

PROOF. Let $\{a_k\}$ be a dense set in E. Define

$$G_{n,k} = \bigcup_{j=1}^{n} B\left(a_j, \frac{1}{k}\right).$$

The maps $\mu \to \mu(G_{n,k})$ are lower semicontinuous and for each k, $\mu(G_{n,k}) \uparrow 1$. By Dini's theorem for each $\epsilon > 0$ and k there exists an n_k such that for all $\mu \in \Gamma$

$$\mu(G_{n_k, k}) \ge 1 - \frac{\epsilon}{2^k}.$$

Define $K = \bigcap_{k=1}^{\infty} \overline{G_{n_k,k}}$. Then for all $\mu \in \Gamma$

$$\mu(K) = \mu\left(\bigcap_{k=1}^{\infty} \overline{G_{n_k,k}}\right) = 1 - \mu\left(\left(\bigcap_{k=1}^{\infty} \overline{G_{n_k,k}}\right)^c\right) = 1 - \mu\left(\bigcup_{k=1}^{\infty} \overline{G_{n_k,k}}^c\right)$$

$$\geq 1 - \sum_{k=1}^{\infty} \mu(\overline{G_{n_k,k}}^c),$$

$$\mu(\overline{G_{n_k,k}}^c) = 1 - \mu(\overline{G_{n_k,k}}) \leq \frac{\epsilon}{2^k}.$$

Consequently $\mu(K) \geq 1 - \epsilon$. The set is closed, and totally bounded because for each k

$$K \subset \bigcup_{j=1}^{n_k} B\left(a_j, \frac{2}{k}\right).$$

LEMMA 10.4.2. *If a metric space E is compact then it is complete and totally bounded. Conversely, a complete and totally bounded space is compact.*

This completes the proof of (i).
To show (ii) we use Riesz's theorem.

THEOREM 10.4.3. *If a metric space K is compact then an arbitrary linear functional ψ on $C(K)$ such that $\psi(f) \geq 0$ for all $f \geq 0$ is of the form*

$$\psi(f) = \int_E f(x) \mu(dx)$$

where μ is a nonnegative finite measure and the representation is unique.

Proof of (ii). Assume that E is compact and define $\psi_\mu(f) = \int_E f(x) \mu(dx)$. Note that $C(E)$ is a separable space, by Lemma 10.3.2, and therefore, given an arbitrary sequence in Γ, by a diagonal procedure one can extract a subsequence ψ_{μ_n} such that $\psi_{\mu_n}(f)$ is convergent for all f in a dense subset of $C(E)$. But then ψ_{μ_n} is convergent (by 3ϵ-method) on the whole $C(E)$ and its limit is a nonnegative functional on $C(E)$. By Riesz's theorem the result follows if E is compact.

Next assume that E is σ-compact, i.e. a countable union of compact subsets. We saw in the proof of Lemma 10.3.1 that there exists a homeomorphism h of E onto a subset $h(E)$ of a compact metric space F (and one can take $F = [0,1]^{\mathbb{N}}$). Let us identify E with $h(E)$ and think of E as a topological subspace of F. Since compact sets of E are compact subsets of F as well, it follows that E is a σ-compact subset of F, hence a Borel subset of F. Thus $\mathcal{M}_1(F) \supseteq \mathcal{M}_1(E)$. Assume that $(\mu_n) \subset \Gamma$. Then there exists a subsequence (μ_{n_k}) such that $\mu_{n_k} \Rightarrow \mu$ in $\mathcal{M}_1(F)$. For arbitrary $\epsilon > 0$ there exists a compact subset $K_\epsilon \subset E$ such that

$$\mu_{n_k}(K_\epsilon) \geq 1 - \epsilon \quad \text{for all } k = 1, 2, \ldots .$$

Since K_ϵ is closed in F as well therefore

$$\overline{\lim_k} \, \mu_{n_k}(K_\epsilon) \leq \mu(K_\epsilon)$$

and we see that $\mu(K_\epsilon) \geq 1 - \epsilon$. Thus $\mu(F\backslash E) = 0$. So μ is supported by E. An arbitrary $f \in UC_b(E)$ can be uniquely extended to $f \in C(F)$ and

$$\int_E f(x)\,\mu_{n_k}(dx) = \int_F f(x)\,\mu_{n_k}(dx) \to \int_F f(x)\,\mu(dx) = \int_E f(x)\,\mu(dx).$$

This proves the result if E is σ-compact.

In the general case, it follows from the tightness condition that there exists a σ-compact subset E_0 of E such that $\mu(E\backslash E_0) = 0$ for all $\mu \in \Gamma$. We can replace E by E_0 and the general case follows from the result proved above. This completes the proof. □

CHAPTER 11

Invariance principle and Kolmogorov's test

We develop some criteria of weak convergence and tightness in the space $C([0, T); E)$ of continuous functions with values in a metric space E The convergence of random walks to the Wiener process is given as application. We introduce also the factorisation method, due to L. Schwartz, to establish tightness and prove Donsker's invariance principle. Finally we prove Kolmogorov's continuity criteria in the context of weak convergence.

11.1. – Weak convergence in $C([0, T]; E)$

Let (E, ρ) be a complete, separable metric space and $C([0, T]; E)$ the space of all continuous functions defined on $[0, T]$ with values in E equipped with metric R:

$$(11.1.1) \qquad R(x, y) = \sup_{t \in [0,T]} \rho(x(t), y(t)).$$

The space $C([0, T]; E)$ is also a complete metric space. To find out whether a given sequence of measures on $C([0, T]; E)$ is tight it is necessary to know characterisations of compact subsets of $C([0, T]; E)$. We have the following generalisation of the Arzelà-Ascoli theorem.

PROPOSITION 11.1.1. *A set* $K \subset C([0, T]; E)$ *has a compact closure in* $C([0, T]; E)$ *if and only if*

(*i*) *There exists a compact set* $L \subset E$ *such that*

$$(11.1.2) \qquad K \subset \{x \,; \, x(t) \in L \text{ for all } t \in [0, T]\}.$$

(ii) *For arbitrary $\epsilon > 0$ there exists $\delta > 0$ such that*

(11.1.3) $$K \subset \{x \; ; \; \sup_{\substack{|t-s| \leq \delta \\ t,s \in [0,T]}} \rho(x(t), x(s)) \leq \epsilon\}.$$

For each $x \in C([0, T]; E)$ define the modulus of continuity $\omega_\delta(x)$, $\delta > 0$, by the formula

(11.1.4) $$\omega_\delta(x) = \sup_{\substack{|t-s| \leq \delta \\ t,s \in [0,T]}} \rho(x(t), y(t)), \qquad \delta > 0.$$

Then the inclusion (11.1.3) can be written

$$K \subset \{x \; ; \; \omega_\delta(x) \leq \epsilon\}.$$

THEOREM 11.1.2. *A sequence (μ_n) of probability measures on $C([0, T]; E)$ is tight if and only if*

(i) *For arbitrary $\epsilon > 0$ there exists a compact set $L \subset E$ such that for all $n = 1, 2, \ldots$*

(11.1.5) $$\mu_n(x \; ; \; x(t) \in L \text{ for all } t \in [0, T]) \geq 1 - \epsilon.$$

(ii) *For arbitrary $\epsilon > 0$ and arbitrary $\eta > 0$ there exists $\delta > 0$ such that for all $n = 1, 2, \ldots$*

(11.1.6) $$\mu_n(x \; ; \; \omega_\delta(x) \leq \eta) \geq 1 - \epsilon.$$

PROOF. Assume that (μ_n) is tight and let K_ϵ be a compact set such that

$$\mu_n(K_\epsilon) \geq 1 - \epsilon, \qquad \text{for all } n = 1, 2, \ldots .$$

Since K_ϵ is compact in $C([0, T]; E)$, there exists a compact set $L_\epsilon \subset E$ such that

$$K_\epsilon \subset \{x \; ; \; x(t) \in L_\epsilon \text{ for all } t \in [0, T]\}.$$

Therefore

$$\mu_n(x \; ; \; x(t) \in L_\epsilon \text{ for all } t \in [0, T]) \geq 1 - \epsilon \qquad \text{for all } n = 1, 2, \ldots .$$

In a similar way, by the compactness of K_ϵ, for arbitrary $\eta > 0$ there exists $\delta > 0$ such that if $x \in K_\epsilon$ then $\omega_\delta(x) \leq \eta$. Therefore for each $n \in \mathbb{N}$,

$$\mu_n(x \; ; \; \omega_\delta(x) \leq \eta) \geq \mu_n(K_\epsilon) \geq 1 - \epsilon.$$

We show now that conditions (i), (ii) imply tightness of (μ_n). Fix $\epsilon > 0$ and choose first a compact set $L_\epsilon \subset E$ such that (11.1.5) holds for all n with ϵ replaced by $\epsilon/2$. Let (η_m) be a sequence of positive numbers converging to zero. There exists a sequence of numbers $\delta_m > 0$ such that for all n, and all m:

$$\mu_n(x \; ; \; \omega_{\delta_m}(x) > \eta_m) < \frac{\epsilon}{2^{m+1}}.$$

Define

$$K_\epsilon = \{x \; ; \; x(t) \in L_\epsilon \text{ for all } t \in [0, T]\} \cap \bigcap_{m=1}^{\infty} \{x \; ; \; \omega_{\delta_m}(x) \leq \eta_m\} = B_0 \cap \bigcap_{m=1}^{\infty} B_m.$$

It is easy to see that K_ϵ is closed and by the Arzelà-Ascoli theorem it is compact. Now for each n:

$$\mu_n(K_\epsilon) = 1 - \mu_n\left(B_0^c \cup \bigcup_{m=1}^{\infty} B_m^c\right) \geq 1 - \frac{\epsilon}{2} - \sum_{m=1}^{\infty} \mu_n(B_m^c)$$

$$\geq 1 - \frac{\epsilon}{2} - \sum_{m=1}^{\infty} \frac{\epsilon}{2^{m+1}} \geq 1 - \epsilon,$$

as required.   □

REMARK 11.1.3.   Condition (i) in Theorem 11.1.2 is equivalent to the requirement that there exists an increasing sequence of compact sets $L_m \subset E$ such that

(11.1.7) $\lim_m \sup_n \mu_n(x \; ; \; x(t) \in L_m \text{ for all } t \in [0, T]) = 0.$

Similarly condition (ii) can be replaced by the following one:

(11.1.8)   For arbitrary $\eta > 0$, $\lim_{\delta \to 0} \left[\overline{\lim_n} \, \mu_n(x \; ; \; \omega_\delta(x) > \eta) \right] = 0.$

We show for instance that (11.1.8) implies (11.1.6). If (11.1.8) holds then for $\epsilon > 0$ there exists $\delta_0 > 0$ such that for all $\delta \in (0, \delta_0]$:

$$\overline{\lim_n} \, \mu_n(x \; ; \; \omega_\delta(x) > \eta) < \epsilon.$$

But then there exists n_0 such that for all $n \geq n_0$,

$$\mu_n(x \; ; \; \omega_{\delta_0}(x) > \eta) < \epsilon.$$

Since, by Ulam's theorem, an arbitrary finite set of measures if tight, one can find $\delta_1 < \delta_0$ such that for $n \leq n_0$

$$\mu_n(x \; ; \; \omega_{\delta_1}(x) > \eta) < \epsilon,$$

consequently for all $n = 1, 2, \ldots$

$$\mu_n(x \, ; \, \omega_{\delta_1}(x) > \eta) < \epsilon,$$

as required.

REMARK 11.1.4.   Assume that $E = \mathbb{R}^1$. Then condition (i) in Theorem 11.1.2 can be replaced by the requirement that for arbitrary $\epsilon > 0$ there exists a compact set $L \subset E$ such that for all $n = 1, 2, \ldots$

$$\mu_n(x \, ; \, x(0) \in L) \geq 1 - \epsilon.$$

Indeed, it is easy to check that, together with condition (ii), this implies that (11.1.5) holds.

If X_n, $n = 1, 2, \ldots$ and X are random variables with values in a metric space E and defined on possibly different probability spaces $(\Omega_n, \mathcal{F}_n, \mathbb{P}_n)$, $n = 1, 2, \ldots$, $(\Omega_0, \mathcal{F}_0, \mathbb{P}_0)$ respectively then we say that (X_n) converges weakly to X if and only if $\mathcal{L}(X_n) \Rightarrow \mathcal{L}(X)$ where $\mathcal{L}(Z)$ denotes the distribution of the random variable Z. If (X_n) converges weakly to X then we write $(X_n) \Rightarrow X$.

It is instructive at this moment to recall the so called Skorokhod imbedding theorem although we will not need it.

THEOREM 11.1.5 (Skorokhod).   *Assume that E is a complete, separable metric space. If $(X_n) \Rightarrow X$ then there exists a probability space $(\Omega, \mathcal{F}, \mathbb{P})$ and E-valued random variables X'_n, $n = 1, 2, \ldots$, X' defined on $(\Omega, \mathcal{F}, \mathbb{P})$ such that*

(i) $\mathcal{L}(X'_n) = \mathcal{L}(X_n)$, $\mathcal{L}(X') = \mathcal{L}(X)$.
(ii) *For almost all $\omega \in \Omega$, $X'_n(\omega) \to X(\omega)$ as $n \to \infty$.*

REMARK 11.1.6.   It is clear that if (ii) holds then $(X'_n) \Rightarrow X'$.

We will need the following elementary properties to identify the limiting measure.

PROPOSITION 11.1.7.   *Assume that E is a separable normed space and (X_n) are E-valued random variables. If $X_n \Rightarrow X$ and*

$$X_n = Y_n + \xi_n$$

where $\xi_n \to 0$ in probability then

$$Y_n \Rightarrow X.$$

PROOF.   Let $\phi \in UC_b(E)$. Then $\mathbb{E}(\phi(Y_n)) = \mathbb{E}(\phi(X_n - \xi_n))$ and

$$|\mathbb{E}(\phi(Y_n)) - \mathbb{E}(\phi(X_n))| \leq \mathbb{E}|\phi(X_n - \xi_n) - \phi(X_n)|.$$

For arbitrary $\epsilon > 0$ there exists $\delta > 0$ such that if $\|x - y\| \leq \delta$ then $|\phi(x) - \phi(y)| \leq \epsilon$. Therefore

$$\mathbb{E}\,|\phi(X_n - \xi_n) - \phi(X_n)| \leq \mathbb{E}(\chi_{\|\xi_n\| \leq \delta} |\phi(X_n - \xi_n) - \phi(X_n)|) + 2\|\phi\|\,\mathbb{P}(\|\xi_n\| > \delta)$$
$$\leq \epsilon + 2\|\phi\|\,\mathbb{P}(\|\xi_n\| > \delta)$$
$$\leq 2\epsilon$$

if n is sufficiently large. $\square$

PROPOSITION 11.1.8. *If F is a continuous mapping from a metric space E into a metric space E_1 and E-valued random variables (X_n) converge weakly to X then the random variables $F(X_n)$ converge weakly to $F(X)$.*

PROOF. If $\phi : E_1 \to \mathbb{R}^1$ is bounded and continuous then obviously $\phi(F)$ is a bounded and continuous function from E into $\mathbb{R}^1$ and the result follows. □

THEOREM 11.1.9. *Let E be a Polish space. A sequence of probability measures (μ_m) on $C([0, T]; E)$ converges weakly to a measure μ if and only if*

(*i*) *(μ_n) is tight.*

(*ii*) *For any sequence $0 = t_0 \leq t_1 \leq t_2 \leq \ldots \leq t_k \leq T$ the finite dimensional distributions $\mu_m^{(t_1,\ldots,t_k)}$ converge weakly to $\mu^{(t_1,\ldots,t_k)}$.*

REMARK 11.1.10. As we already know, if μ is a measure on $C([0, T]; E)$ then $\mu^{(t_1,\ldots,t_k)}$ is a measure on E^k given by the formula

$$\mu^{(t_1,\ldots,t_k)}(\Gamma) = \mu\{x \, ; \, (x(t_1), \ldots, x(t_k)) \in \Gamma\} \qquad \text{for any } \Gamma \in \mathcal{B}(E^k).$$

PROOF. If $\mu_m \Rightarrow \mu$ then (μ_m) is tight so (*i*) holds. If $\psi \in C_b(E^k)$ then

$$\phi(x) = \psi(x(t_1), \ldots, x(t_k)), \qquad x \in C([0, T]; E),$$

is a bounded continuous function on $C([0, T]; E)$ and therefore

$$\int_{C([0,T];E)} \phi(x)\, \mu_m(dx) \to \int_{C([0,T];E)} \phi(x)\, \mu(dx).$$

But

$$\int_{C([0,T];E)} \phi(x)\, \mu_m(dx) = \int_{E^k} \psi(y)\, \mu_m^{(t_1,\ldots,t_k)}(dy),$$

so $\mu_m^{(t_1,\ldots,t_k)} \Rightarrow \mu^{(t_1,\ldots,t_k)}$ on E^k.

Conversely, assume that (μ_n) is tight and (*ii*) holds. It is enough to show that arbitrary weakly convergent subsequences of (μ_m) converge to the same limit. If μ^1 and μ^2 are two such limits then, for arbitrary $k = 1, 2, \ldots$ and $\Gamma \in \mathcal{B}(E^k)$,

$$\mu^1(x \, ; \, (x(t_1), \ldots, x(t_k)) \in \Gamma) = \mu^2(x \, ; \, (x(t_1), \ldots, x(t_k)) \in \Gamma).$$

But the cylindrical sets $\{x \, ; \, (x(t_1), \ldots, x(t_k)) \in \Gamma\}$ generate the Borel σ-field $\mathcal{B}(C([0, T]; E))$ so by Dynkin's $\pi - \lambda$ theorem, $\mu^1 = \mu^2$. □

11.2. – Classical proof of the invariance principle

Let ξ_n^m be independent real valued random variables defined on a probability space $(\Omega, \mathcal{F}, \mathbb{P})$,

$$\mathbb{E}\,\xi_n^m = 0, \quad \mathbb{E}\,(\xi_n^m)^2 = \frac{1}{m}, \quad n, m = 1, 2, \ldots,$$

which, in addition, have Gaussian distributions. We define, by induction, the sequences $(B_n^m)_{n=1,\ldots}$:

$$B_{n+1}^m = B_n^m + \xi_{n+1}^m, \quad B_0^m = 0, \quad n = 0, 1, 2, \ldots.$$

Let (X_t^m) be the following stochastic processes:

$$(11.2.1) \quad X_t^m = B_n^m + (mt - n)(B_{n+1}^m - B_n^m) = B_n^m + (mt - n)\xi_{n+1}^m, \quad \text{for } t \in \left[\frac{n}{m}, \frac{n+1}{m}\right].$$

Let us recall that a *Wiener process* is a stochastic process W such that: $i)$ $W(0) = 0$; $ii)$ for arbitrary $0 = t_0 < t_1 < \ldots < t_k$ the random variables

$$W(t_1), W(t_2) - W(t_1), \ldots, W(t_k) - W(t_{k-1})$$

are centred Gaussian and independent; $iii)$ $\mathbb{E}|W(t) - W(s)|^2 = t - s, \ t \geq s \geq 0$; $iv)$ the trajectories of W are continuous with probability one.

On the measurable space $(E, \mathcal{B}(E))$ where $E = C([0, T])$ define a canonical process X as follows:

$$X_t(f) = f(t), \quad f \in E, t \in [0, T].$$

THEOREM 11.2.1. *The distributions μ_m of the processes X^m converge weakly to a measure μ and the canonical process on $(E, \mathcal{B}(E), \mu)$ is a Wiener process. Consequently the Wiener process W exists and $X_m \Rightarrow W$.*

PROOF. STEP 1. We will show first that the sequence (μ_m) is tight on E. Since for all $m = 1, 2, \ldots, X_0^m = 0$ it is enough to prove that

$$(11.2.2) \qquad \lim_{\delta \to 0} \overline{\lim_{m}} \ \mathbb{P}\left(\sup_{\substack{|t-s| \leq \delta \\ t,s \in [0,T]}} |X_t^m - X_s^m| > \epsilon\right) = 0$$

for arbitrary fixed $\epsilon > 0$. Choose $\delta \in (0, T]$ and assume that $1/m < \delta$. Define

$$k_m = \max\left\{k : \frac{k}{m} \leq \delta\right\} + 1, \quad k_m = [m\delta] + 1$$

$$l_m = \inf\left\{l; l\frac{k_m}{m} \geq T\right\} - 1, \quad m = 1, 2, \ldots.$$

Thus k_m denotes the minimal number of intervals of length $1/m$ covering the interval $[0, \delta]$ and l_m measures how many nonintersecting intervals of length k_m/m (equal approximately to δ) can be included in $[0, T]$. Let

$$I_l^m = \left[l \frac{k_m}{m}, (l+1) \frac{k_m}{m} \right], \qquad l = 0, 1, 2, \ldots .$$

We show first that

(11.2.3) $$\sup_{\substack{|t-s| \leq \delta \\ t,s \in [0,T]}} |X_t^m - X_s^m| \leq 3 \max_{0 \leq l \leq l_m - 1} \left[\max_{lk_m < j \leq (l+1)k_m} |B_j^m - B_{lk_m}^m| \right].$$

If $|t - s| \leq \delta$ and $t, s \in I_l^m$ for some l then

$$|X_t^m - X_s^m| \leq |X_{j \frac{k_m}{m}}^m - X_{k \frac{k_m}{m}}^m|$$

for some $j, k \in \{lk_m, lk_m + 1, \ldots, (l+1)k_m\}$, and therefore

$$|X_t^m - X_s^m| \leq |X_{j \frac{k_m}{m}}^m - X_{l \frac{k_m}{m}}^m| + |X_{k \frac{k_m}{m}}^m - X_{l \frac{k_m}{m}}^m|$$
$$\leq |B_{jk_m}^m - B_{lk_m}^m| + \| B_{kk_m}^m - B_{lk_m}^m|.$$

If $|t - s| \leq \delta$ and $t \in I_l^m$, $s \in I_{l+1}^m$ then in a similar way:

$$|X_t^m - X_s^m| \leq |X_{j \frac{k_m}{m}}^m - X_{k \frac{k_m}{m}}^m|$$

where $j \in \{lk_m, \ldots, (l+1)k_m\}$, $k \in \{(l+1)k_m, \ldots, (l+2)k_m\}$, and

$$|X_t^m - X_s^m| \leq |X_{j \frac{k_m}{m}}^m - X_{l \frac{k_m}{m}}^m| + |X_{(l+1) \frac{k_m}{m}}^m - X_{l \frac{k_m}{m}}^m| + |X_{k \frac{k_m}{m}}^m - X_{(l+1) \frac{k_m}{m}}^m|$$
$$\leq |B_{jk_m}^m - B_{lk_m}^m| + |B_{(l+1)k_m}^m - B_{lk_m}^m| + |B_{kk_m}^m - B_{(l+1)k_m}^m|.$$

This way the estimate (11.2.3) has been proved.

It follows from (11.2.3) that

$$\mathbb{P} \left(\sup_{\substack{|t-s| \leq \delta \\ t,s \in [0,T]}} |X_t^m - X_s^m| > \epsilon \right) \leq \mathbb{P} \left(\max_{0 \leq l \leq l_m - 1} \max_{lk_m < j \leq (l+1)k_m} |B_j^m - B_{lk_m}^m| > \frac{\epsilon}{3} \right)$$

$$\leq \sum_{l=0}^{l_m - 1} \mathbb{P} \left(\max_{lk_m < j \leq (l+1)k_m} |B_j^m - B_{lk_m}^m| > \frac{\epsilon}{3} \right).$$

By Ottaviani's inequality, see Subsection 11.3,

$$\mathbb{P} \left(\max_{lk_m < j \leq (l+1)k_m} |B_j^m - B_{lk_m}^m| > 2\frac{\epsilon}{6} \right) \left(1 - \max_{lk_m < j \leq (l+1)k_m} \mathbb{P} \left(|B_{(l+1)k_m}^m - B_j^m| > \frac{\epsilon}{6} \right) \right)$$

$$\leq \mathbb{P} \left(|B_{(l+1)k_m}^m - B_{lk_m}^m| > \frac{\epsilon}{3} \right).$$

However by Chebyshev's inequality and the definition of B_j^m:

$$\mathbb{P}\left(|B_{(l+1)k_m}^m - B_{lk_m}^m| > \frac{\epsilon}{6}\right) \leq \frac{\mathbb{E}|B_{(l+1)k_m}^m - B_{lk_m}^m|^2}{\left(\frac{\epsilon}{6}\right)^2} \leq \frac{k_m}{m}\frac{36}{\epsilon^2}.$$

Consequently,

$$\mathbb{P}\left(\max_{lk_m < j \leq (l+1)k_m} |B_j^m - B_{lk_m}^m| > \frac{\epsilon}{3}\right) \leq \frac{1}{1 - \frac{k_m}{m}\frac{36}{\epsilon^2}}\mathbb{P}\left(|B_{k_m}^m| > \frac{\epsilon}{3}\right).$$

Taking into account that $\frac{k_m}{m} \to \delta$, $1 - \frac{k_m}{m}\frac{36}{\epsilon^2} \to 1 - \delta\frac{36}{\epsilon^2}$ and that $l_m \leq \frac{T}{\delta}+1 \leq \frac{2T}{\delta}$ we have that

$$\overline{\lim_m}\,\mathbb{P}\left(\sup_{\substack{|t-s|\leq\delta \\ t,s\in[0,T]}} |X_t^m - X_s^m|\right) \leq \overline{\lim_m}\,\frac{l_m}{1 - \frac{k_m}{m}\frac{36}{\epsilon^2}}\mathbb{P}\left(|B_{k_m}^m| > \frac{\epsilon}{3}\right) \leq \frac{2T\frac{1}{\delta}\mathbb{P}\left(|\zeta_\delta| > \frac{\epsilon}{3}\right)}{1 - \delta\frac{36}{\epsilon^2}},$$

where $\mathcal{L}(\zeta_\delta) = N(0, \delta)$. It is therefore enough to show that

(11.2.4) $$\lim_{\delta\to 0}\frac{1}{\delta}\frac{1}{\sqrt{2\pi\delta}}\int_{|x|>\frac{\epsilon}{3}} e^{-\frac{|x|^2}{2\delta}}\,dx = 0.$$

However, for any $c > 0$, taking $\frac{x}{\sqrt{\delta}} = y$,

$$\frac{1}{\sqrt{2\pi\delta}}\int_{|x|>c} e^{-\frac{|x|^2}{2\delta}}\,dx = \frac{1}{\sqrt{2\pi}}\int_{|y|>\frac{c}{\sqrt{\delta}}} e^{-\frac{|y|^2}{2}}\,dy$$

$$\leq \frac{1}{\sqrt{2\pi}}\int_{|y|>\frac{c}{\sqrt{\delta}}} \left(|y|\frac{\sqrt{\delta}}{c}\right)^2 e^{-\frac{|y|^2}{2}}\,dy = \frac{\delta}{c^2\sqrt{2\pi}}\int_{|y|>\frac{c}{\sqrt{\delta}}} y^2 e^{-\frac{|y|^2}{2}}\,dy.$$

Since

$$\lim_{\delta\to 0}\int_{|y|>\frac{c}{\sqrt{\delta}}} y^2 e^{-\frac{|y|^2}{2}}\,dy = 0,$$

the identity (11.2.4) holds.

STEP 2. We show that the finite dimensional distributions of X^m converge weakly to the finite dimensional distributions of W. Although we do not know whether $\mathcal{L}(W)$ is supported by $C([0, T]; \mathbb{R}^1)$, the statement is meaningful.

Let $0 = t_0 < t_1 < t_2 < \ldots < t_k \leq T$. It is enough to show that the laws of

$$(X_{t_1}^m, X_{t_2}^m - X_{t_1}^m, \ldots, X_{t_k}^m - X_{t_{k-1}}^m)$$

converge to the law of

$$(W(t_1), W(t_2) - W(t_1), \ldots, W(t_k) - W(t_{k-1})).$$

For each $m = 1, 2, \ldots, j = 1, 2, \ldots$ set

$$t_{j,m} = \max\left\{\frac{k}{m}; \frac{k}{m} \leq t_j\right\}, \qquad j = 1, \ldots, k.$$

It is enough to show that the random variables

$$(X^m_{t_{1,m}}, X^m_{t_{2,m}} - X^m_{t_{1,m}}, \ldots, X^m_{t_{k,m}} - X^m_{t_{k-1,m}})$$

converge weakly to

$$(W(t_1), W(t_2) - W(t_1), \ldots, W(t_k) - W(t_{k-1})).$$

[1] However $X^m_{t_{1,m}}, X^m_{t_{2,m}} - X^m_{t_{1,m}}, \ldots, X^m_{t_{k,m}} - X^m_{t_{k-1,m}}$ are independent Gaussian random variables identical with

$$(B^m_{mt_{1,m}}, B^m_{mt_{2,m}} - B^m_{mt_{1,m}}, \ldots, B^m_{mt_{k,m}} - B^m_{mt_{k-1,m}}).$$

The second moments of these random variables are

$$t_{1,m}, t_{2,m} - t_{1,m}, \ldots, t_{k,m} - t_{k-1,m}$$

and they converge to

$$t_1, t_2 - t_1, \ldots, t_k - t_{k-1}.$$

This easily implies the required convergence. □

[1] Indeed we have, for instance,

$$X^m_{t_1} - X^m_{t_{1,m}} = \theta^{(1)}_m \xi^{(1)}_m,$$

with $\theta^{(1)}_m$ a number in $[0, 1]$ and $\xi^{(1)}_m$ a random variable with $\mathbb{E}(\xi^{(1)}_m)^2 = 1/m$, hence converging to zero in probability; by Proposition 11.1.7 the random variables $X^m_{t_1}$ and $X^m_{t_{1,m}}$ have the same weak limit.

11.3. – The Ottaviani inequality

THEOREM 11.3.1. *Let* $\xi_1, \xi_2, \ldots$ *be independent random variables with values in a normed space* $(E, | \cdot |)$. *Define* $S_k = \xi_1 + \ldots + \xi_k$, $k = 1, 2 \ldots$. *Then for arbitrary* $r \geq 0$ *and arbitrary natural numbers* $n > m \geq 1$

$$\mathbb{P}\left(\max_{m < j \leq n} |S_j - S_m| > 2r \right) \left(1 - \max_{m < j \leq n} \mathbb{P}(|S_n - S_j| > r) \right) \leq \mathbb{P}(|S_n - S_m| > r).$$

PROOF. Let

$$\tau = \min (k \in \{m + 1, \ldots, n\}; |S_k - S_m| > 2r)$$
$$= +\infty \text{ if } |S_k - S_m| \leq 2r \text{ for } k = m + 1, \ldots, n.$$

Note that if $\tau = k$ and $|S_n - S_k| \leq r$ then

$$|S_n - S_m| \geq |S_k - S_m| - |S_n - S_k| > r, \qquad k = m + 1, \ldots, n.$$

Consequently

(11.3.1) $\mathbb{P}(\tau = k \text{ and } |S_n - S_k| \leq r) \leq \mathbb{P}(\tau = k \text{ and } |S_n - S_m| > r).$

The event $\{\tau = k\}$ is $\sigma\{\xi_{m+1}, \ldots, \xi_k\}$ measurable and the event $\{|S_n - S_k| \leq r\}$ is $\sigma\{\xi_{k+1}, \ldots, \xi_n\}$ measurable so they are independent. Therefore

$$\mathbb{P}(\tau = k \text{ and } |S_n - S_k| \leq r) = \mathbb{P}(\tau = k) \mathbb{P}(|S_n - S_k| \leq r)$$

and by (11.3.1)

(11.3.2) $\mathbb{P}(\tau = k) \mathbb{P}(|S_n - S_k| \leq r) \leq \mathbb{P}(\tau = k \text{ and } |S_n - S_m| > r).$

Adding inequalities in (11.3.2) with respect to $k = m + 1, \ldots, n$ one gets

$$\left(\sum_{k=m+1}^{n} \mathbb{P}(\tau = k) \right) \min_{m+1 \leq k \leq n} \mathbb{P}(|S_n - S_k| \leq r) \leq \mathbb{P}(|S_n - S_m| > r).$$

But $\sum_{k=m+1}^{n} \mathbb{P}(\tau = k) = \mathbb{P}(\max_{m < j \leq n} |S_j - S_m| > 2r)$ and the result follows. $\square$

11.4. – Tightness by the factorisation method

We prove now the tightness of the sequence (μ_m) using the so called factorisation method which goes back to L. Schwartz, see a discussion in [62]. Both methods, the classical one and the factorisation, have much wider range of applications.

For arbitrary integrable function f defined on $[0, T]$ denote by $I_1 f$ its integral:

$$(11.4.1) \qquad I_1 f(t) = \int_0^t f(s)\,ds, \qquad t \in [0, T],$$

and, more generally, by $I_\alpha f$, $\alpha > 0$, the α-fractional integral of f, whenever defined:

$$(11.4.2) \qquad I_\alpha f(t) = \frac{1}{\Gamma(\alpha)} \int_0^t (t - s)^{\alpha-1} f(s)\,ds, \qquad t \in [0, T].$$

The α-fractional integral of f is called also the Riemann-Liouville integral of f, see [50] The operators I_α, $\alpha > 0$, have the semigroup property

$$I_{\alpha+\beta} f = I_\alpha(I_\beta f)$$

and in particular, for $\alpha \in (0, 1)$,

$$(11.4.3) \qquad I_1 f = I_\alpha(I_{1-\alpha} f).$$

To check (11.4.3) notice that

$$I_\alpha(I_{1-\alpha} f)(t) = \frac{1}{\Gamma(\alpha)} \frac{1}{\Gamma(1-\alpha)} \int_0^t (t - s)^{\alpha-1} \left[\int_0^s (s - \sigma)^{-\alpha} f(\sigma)\,d\sigma \right] ds$$

$$= \frac{1}{\Gamma(\alpha)\,\Gamma(1-\alpha)} \int_0^t \left[\int_\sigma^t (t - s)^{\alpha-1}(s - \sigma)^{-\alpha}\,ds \right] f(\sigma)\,d\sigma.$$

But changing variables $u = (t - \sigma)v$

$$\int_\sigma^t (t - s)^{\alpha-1}(s - \sigma)^{-\alpha}\,ds = \int_0^{t-\sigma} (t - \sigma - u)^{\alpha-1} u^{-\alpha}\,du$$

$$= \int_0^1 (t - \sigma)^{\alpha-1}(1 - v)^{\alpha-1}(t - \sigma)^{-\alpha} v^{-\alpha}(t - \sigma)\,dv = \int_0^1 (1 - v)^{\alpha-1} v^{-\alpha}\,dv.$$

We will call (11.4.3) the factorisation formula. For its infinite dimensional generalisation see [13] and [14]. We will need the following functional analytic result:

PROPOSITION 11.4.1. *If* $1/p < \alpha \le 1$ *then the operators* I_α *are compact from* $L^p[0, T]$ *into* $C[0, T]$.

THEOREM 11.4.2. *The sequence* (μ_m) *is tight on* $C[0, T]$.

PROOF. Note first that

$$(11.4.4) \qquad X_t^m = \int_0^t \dot{X}_s^m \, ds = I_1(\dot{X}^m)(t) = I_\alpha(I_{1-\alpha}(\dot{X}^m))(t),$$

where we denoted by $\dot{X}_s^m$ the derivative $\frac{d}{ds} X_s^m$ (which is a piecewise constant function for each ω). One can treat the random processes $I_{1-\alpha}(\dot{X}^m)$ as $L^p[0, T]$-valued random variables. Taking into account Proposition 11.4.1 it is enough to show that

LEMMA 11.4.3. *If* $2\alpha < 1$ *then for all* $m = 1, \dots$ *and* $p > 0$

$$(11.4.5) \qquad \mathbb{E}\|I_{1-\alpha}(\dot{X}^m)\|_{L^p[0,T]} \le \frac{c_p}{\Gamma^p(1-\alpha)} \frac{T^{1+(1-2\alpha)(p/2)}}{(1-2\alpha)^{p/2}},$$

where c_p *is the p-th moment of a normalised Gaussian random variable.*

In fact by Chebyshev's inequality, for arbitrary $r > 0$,

$$\mathbb{P}(\|I_{1-\alpha}(\dot{X}^m)\|_{L^p[0,T]} \ge r) \le \frac{\mathbb{E}\|I_{1-\alpha}(\dot{X}^m)\|_{L^p[0,T]}^p}{r^p},$$

so by (11.4.5)

$$\lim_{r \to \infty} \left[\sup_m \mathbb{P}\left(\|I_{1-\alpha}(\dot{X}^m)\|_{L^p[0,T]} > r \right) \right] = 0.$$

If p and α are chosen in such a way that

$$\frac{1}{p} < \alpha < \frac{1}{2}$$

then by Proposition 11.4.1 and by the semigroup property (11.4.3), the laws $\mathcal{L}(X^m)$ are tight on $C[0, T]$.

It remains to prove Lemma 11.4.3.

Note that, by the very definition,

$$\dot{X}_\sigma^m = m \, \xi_{n+1}^m \qquad \text{for} \quad \frac{n}{m} < \sigma < \frac{n+1}{m}.$$

Consequently,

$$\dot{X}_\sigma^m = m \sum_{n=0}^\infty \chi_{(\frac{n}{m}, \frac{n+1}{m})}(\sigma) \xi_{n+1}^m, \qquad \sigma \in [0, T],$$

and

$$\mathbb{E}\|I_{1-\alpha}(\dot{X}^m)\|^p_{L^p[0,T]} = m^p\,\mathbb{E}\left[\int_0^T \left(\sum_n (I_{1-\alpha}\chi_{(\frac{n}{m},\frac{n+1}{m})})(t)\xi^m_{n+1}\right)^p dt\right].$$

If ζ is a Gaussian random variable, $\mathbb{E}\zeta = 0$, then $\mathbb{E}|\zeta|^p = c_p(\mathbb{E}|\zeta|^2)^{p/2}$, where c_p is the p-th moment of a normalised Gaussian random variable. Therefore

$$\mathbb{E}\|I_{1-\alpha}(\dot{X}^m)\|^p_{L^p[0,T]} = m^p c_p \int_0^T \left[\mathbb{E}\left(\sum_n (I_{1-\alpha}\chi_{(\frac{n}{m},\frac{n+1}{m})})(t)\xi^m_{n+1}\right)^2\right]^{p/2} dt.$$

But, for each $t \in [0, T]$,

$$\mathbb{E}\left(\sum_n (I_{1-\alpha}\chi_{(\frac{n}{m},\frac{n+1}{m})})(t)\xi^m_{n+1}\right)^2 = \frac{1}{m}\sum_n (I_{1-\alpha}\chi_{(\frac{n}{m},\frac{n+1}{m})})(t)^2$$

and

$$(I_{1-\alpha}\chi_{(\frac{n}{m},\frac{n+1}{m})})(t)^2 = \frac{1}{\Gamma^2(1-\alpha)}\langle\chi_{[0,t]}(t-\cdot)^{-\alpha},\chi_{(\frac{n}{m},\frac{n+1}{m})}\rangle^2_{L^2[0,T]}.$$

Since

$$\|\chi_{(\frac{n}{m},\frac{n+1}{m})}\|^2_{L^2[0,T]} = \frac{1}{m}$$

and, for different n, $\chi_{(\frac{n}{m},\frac{n+1}{m})}$ are orthogonal functions, consequently, by Parseval's inequality:

$$\sum_n \langle\chi_{[0,t]}(t-\cdot)^{-\alpha},\chi_{(\frac{n}{m},\frac{n+1}{m})}\rangle^2 = \frac{1}{m}\sum_n \langle\chi_{[0,t]}(t-\cdot)^{-\alpha},\sqrt{m}\chi_{(\frac{n}{m},\frac{n+1}{m})}\rangle^2$$

$$\leq \frac{1}{m}\int_0^t (t-\sigma)^{-2\alpha}d\sigma = \frac{1}{m}\int_0^t \sigma^{-2\alpha}d\sigma,\ t \in [0, T].$$

This way we arrive at the estimate:

$$\mathbb{E}\|I_{1-\alpha}(\dot{X}^m)\|^p_{L^p[0,T]} \leq m^p c_p \frac{1}{m^{p/2}}\frac{1}{\Gamma^p(1-\alpha)}\int_0^T \left(\frac{1}{m}\int_0^T \sigma^{-2\alpha}d\sigma\right)^{p/2} dt$$

$$\leq \frac{c_p}{\Gamma^p(1-\alpha)}T\frac{T^{(1-2\alpha)(p/2)}}{(1-2\alpha)^{p/2}},$$

which proves the lemma. $\square$

11.5. – Donsker's theorem

THEOREM 11.5.1. *Assume that for each* m, $\xi_1^m, \xi_2^m, \ldots$ *are independent, identically distributed such that*

$$\mathbb{E}\,\xi_n^m = 0, \quad \mathbb{E}\,(\xi_n^m)^2 = \frac{1}{m}, \qquad m = 1, 2, \ldots, \ n = 1, 2, \ldots,$$

then the processes X^m *converge weakly to a Wiener process.*

PROOF. The proof is the same as for Gaussian ξ_n^m. Only in the final part one has to use the following central limit theorem.

CENTRAL LIMIT THEOREM. *Under the conditions of Theorem 11.5.1*

$$\mathcal{L}(B_m^m) \Rightarrow N(0, 1) \qquad \text{as } m \to \infty.$$

More generally, if $\frac{k_m}{m} \to \sigma^2$ *as* $m \to \infty$ *then*

$$\mathcal{L}(B_{k_m}^m) \Rightarrow N(0, \sigma^2) \qquad \text{as } m \to \infty.$$

11.6. – Tightness by Kolmogorov's test

Let (E, ρ) be a metric space and X_n a sequence of E-valued, continuous processes defined on possibly different probability spaces $(\Omega_n, \mathcal{F}_n, \mathbb{P}^n)$. We have the following version of the Kolmogorov's continuity test, see [56].

THEOREM 11.6.1. *Assume that there exist positive numbers* α, r, c, *such that for all* $s, t \in [0, T]$

$$(11.6.1) \qquad \mathbb{E}^n(\rho^r(X_n(t), X_n(s))) \leq c\,|t - s|^{1+\alpha}.$$

If the laws of $X_n(0)$ *are tight on* E *then also the laws of* $X_n(\cdot)$ *on* $C([0, T]; E)$ *are tight.*

PROOF. To simplify notation we assume that $T = 1$. Let $\mathcal{D}$ be the set of dyadic numbers $t_{mk} = k/2^m$, $k = 0, 1, 2, \ldots, 2^m$, $m = 0, 1, 2, \ldots$. We show that for arbitrary $\epsilon > 0$

$$\lim_{\delta \to 0} \sup_n \mathbb{P}^n \left(\sup_{\substack{t,s \in \mathcal{D} \\ |t-s| \leq \delta}} \rho(X_n(t), X_n(s)) > \epsilon \right) = 0.$$

Let us fix $k \in \mathbb{N}$ and assume that $\frac{1}{2^k} \leq \delta < \frac{1}{2^{k-1}}$. Then

$$\sup_{\substack{t,s \in \mathcal{D} \\ |t-s| \leq \delta}} \rho(X_n(t), X_n(s)) \leq 3 \sup_j \left[\sup_{\frac{j}{2^k} < \frac{l}{2^m} < \frac{j+1}{2^k}} \rho\left(X_n\left(\frac{l}{2^m}\right), X_n\left(\frac{j}{2^k}\right) \right) \right],$$

where $m > k$.

We have the following elementary lemma.

LEMMA 11.6.2. *Any dyadic number* $x = \frac{l}{2^m} < 1$ *where* $l = 0, 1, \ldots 2^m - 1$ *has a unique representation in the form*:

$$x = \sum_{j=1}^{m} \frac{\epsilon_j}{2^j}, \qquad \text{where } \epsilon_j = 0 \text{ or } 1.$$

PROOF OF THE LEMMA. If $m = 0$ and $m = 1$ the result is true. Let the result be true for $m - 1$ and $\frac{l}{2^m} < 1$. If $l = 2l'$ then $\frac{l'}{2^{m-1}} < 1$ and by induction

$$\frac{l}{2^m} = \frac{l'}{2^{m-1}} = \sum_{j=1}^{m-1} \frac{\epsilon_j}{2^j}.$$

If $l = 2l' + 1$ then

$$\frac{l}{2^m} = \frac{l'}{2^{m-1}} + \frac{1}{2^m}$$

and again we have representation. Uniqueness: assume

$$\sum_{j=1}^{m} \frac{\epsilon_j}{2^j} = \sum_{j=1}^{m} \frac{\epsilon_j'}{2^j}.$$

Let k be the smallest value of the index j such that $\epsilon_j \neq \epsilon_j'$. Without loss of generality we may suppose $\epsilon_k = 1$, $\epsilon_k' = 0$. Then

$$\frac{1}{2^k} + \sum_{j=k+1}^{m} \frac{\epsilon_j}{2^j} = \sum_{j=k+1}^{m} \frac{\epsilon_j'}{2^j}.$$

This is impossible, since

$$\frac{1}{2^k} > \sum_{j=k+1}^{m} \frac{1}{2^j} \geq \sum_{j=k+1}^{m} \frac{\epsilon_j'}{2^j}.$$

By the lemma

$$\frac{l}{2^m} = \frac{j}{2^k} + \sum_{r=1}^{s} \frac{1}{2^{m_r}},$$

where $k < m_1 < \ldots < m_s \leq m$. Consequently

$$\rho\left(X_n\left(\frac{l}{2^m}\right), X_n\left(\frac{j}{2^k}\right)\right) \leq \sum_{i=1}^{s} \rho\left(X_n\left(\frac{j}{2^k} + \sum_{r=1}^{i} \frac{1}{2^{m_r}}\right), X_n\left(\frac{1}{2^k} + \sum_{r=1}^{i-1} \frac{1}{2^{m_r}}\right)\right),$$

and

$$\sup_{\substack{t,s \in D \\ |t-s| \leq \delta}} \rho(X_n(t), X_n(s)) \leq 2 \sum_{m=k+1}^{\infty} \left[\sup_{l \leq 2^m - 1} \rho\left(X_n\left(\frac{l+1}{2^m}\right), X_n\left(\frac{l}{2^m}\right)\right) \right].$$

However

$$\mathbb{P}^n \left(\sup_{l \leq 2^m - 1} \rho \left(X_n \left(\frac{l+1}{2^m} \right), X_n \left(\frac{l}{2^m} \right) \right) > \frac{1}{m^2} \right)$$

$$\leq \sum_{l \leq 2^m - 1} \mathbb{P}^n \left(\rho \left(X_n \left(\frac{l+1}{2^m} \right), X_n \left(\frac{l}{2^m} \right) \right) > \frac{1}{m^2} \right)$$

and by Chebyshev's inequality and (11.6.1)

$$\leq m^{2r} \sum_{l \leq 2^m - 1} \mathbb{E}^n \rho^r \left(X_n \left(\frac{l+1}{2^m} \right), X_n \left(\frac{l}{2^m} \right) \right) \leq m^{2r} 2^m c \left(\frac{1}{2^m} \right)^{1+\alpha} = c \frac{m^{2r}}{2^{m\alpha}}.$$

Choose k such that

$$\sum_{m=k+1}^{\infty} \frac{1}{m^2} < \frac{\epsilon}{2},$$

then

$$\mathbb{P}^n \left(\sup_{\substack{t,s \in D \\ |t-s| \leq \delta}} \rho(X_n(t), X_n(s)) > \epsilon \right)$$

$$\leq \mathbb{P}^n \left(2 \sum_{m=k+1}^{\infty} \sup_{l \leq 2^m - 1} \rho \left(X_n \left(\frac{l+1}{2^m} \right), X_n \left(\frac{l}{2^m} \right) \right) > \epsilon \right)$$

$$\leq \mathbb{P}^n \left(\sum_{m=k+1}^{\infty} \sup_{l \leq 2^m - 1} \rho \left(X_n \left(\frac{l+1}{2^m} \right), X_n \left(\frac{l}{2^m} \right) \right) > \sum_{m=k+1}^{\infty} \frac{1}{m^2} \right)$$

$$\leq \sum_{m=k+1}^{\infty} \mathbb{P}^n \left(\sup_{l \leq 2^m - 1} \rho \left(X_n \left(\frac{l+1}{2^m} \right), X_n \left(\frac{l}{2^m} \right) \right) > \frac{1}{m^2} \right)$$

$$\leq c \sum_{m \geq k+1} \frac{m^{2r}}{2^{m\alpha}}.$$

Since the series $\sum_m \frac{m^{2r}}{2^{m\alpha}}$ is convergent, and $\delta \in [\frac{1}{2^k}, \frac{1}{2^{k-1}})$,

$$\sum_{m \geq k+1} \frac{m^{2r}}{2^{m\alpha}} \leq \sum_{m \geq [\log_2 \frac{1}{\delta}]+2} \frac{m^{2r}}{2^{m\alpha}} \to 0 \qquad \text{as } \delta \to 0. \qquad \square$$

Theorem 11.6.1 implies Theorem 11.2.1. This follows from the following facts.

LEMMA 11.6.3. *Assume that $d > 0$ is a positive number and $\xi_1, \xi_2, \ldots$ are independent, real random variables such that for some $\sigma > 0$ and all $k = 1, 2, \ldots,$*

$$\mathbb{E}\,\xi_k = 0, \qquad \mathbb{E}\,\xi_k^2 = \sigma\,d.$$

Define $X_{kd} = \xi_1 + \ldots + \xi_k$, $k = 1, 2, \ldots$ and

$$X_t = X_{kd} + \frac{t - kd}{d}\,\xi_{k+1}, \qquad for\ t \in [kd, (k+1)d].$$

Then

$$\mathbb{E}\,|X_t - X_s|^2 \leq \sigma\,|t - s|, \qquad for\ all\ t, s \geq 0.$$

PROOF. Assume that $s \geq t$ and $s \in [ld, (l+1)d]$. Then

$$X_s - X_t = X_{ld} - X_{kd} + \frac{s - ld}{d}\,\xi_{l+1} - \frac{t - kd}{d}\,\xi_{k+1}$$

$$= \xi_l + \ldots + \xi_{k+1} + \frac{s - ld}{d}\,\xi_{l+1} - \frac{t - kd}{d}\,\xi_{k+1}$$

$$= \frac{s - ld}{d}\,\xi_{l+1} + \xi_l + \xi_{l-1} + \ldots + \xi_{k+2} + \left(1 - \frac{t - kd}{d}\right)\xi_{k+1}.$$

So

$$\mathbb{E}\,|X_t - X_s|^2 = \left(\frac{s - ld}{d}\right)^2 \sigma\,d + \sigma\,d\,(l - (k+1)) + \left(1 - \frac{t - kd}{d}\right)^2 \sigma\,d.$$

Since

$$\sigma\,(s - t) = \sigma\,(s - ld) + \sigma\,d\,(l - (k+1)) + \sigma\,((k+1)d - t),$$

$$\sigma\,d\left(\frac{s - ld}{d}\right)^2 = \frac{s - ld}{d}\sigma\,(s - ld) \leq \sigma\,(s - ld),$$

$$\sigma\,d\left(1 - \frac{t - kd}{d}\right)^2 = \frac{(k+1)d - t}{d}\sigma\,((k+1)d - t) \leq \sigma\,((k+1)d - t),$$

the result follows. $\square$

If ζ is a Gaussian random variable, $\mathbb{E}\,\zeta = 0$, $\mathbb{E}\,\zeta^2 = \gamma^2$, then for arbitrary $p > 0$

$$\mathbb{E}\,|\zeta|^p = \left(\mathbb{E}\,\left|\frac{\zeta}{\gamma}\right|^p\right)\gamma^p$$

and consequently for any $p > 0$ there exists $c_p > 0$ such that

$$\mathbb{E}\,|\zeta|^p \leq c_p\,(\mathbb{E}\,|\zeta|^2)^{p/2}.$$

Thus if in addition to the assumptions of Lemma 11.6.3 one requires that $\xi_1, \ldots$ are Gaussian then

$$\mathbb{E}\,|X_t - X_s|^p \leq c_p\left(\mathbb{E}\,|X_t - X_s|^2\right)^{p/2} \leq c_p\,\sigma^{p/2}|t - s|^{p/2}.$$

So if $p > 2$ the conditions of the Kolomogorov test are satisfied. $\square$

Bibliography

[1] L. BACHELIER, *Théorie de la spéculation*, Ann. Sci École Num. Sup. **17** (1900), 21-86.

[2] J. BERTOIN, *Lévy processes*, Cambridge Univ. Press, 1996.

[3] P. BILLINGSLEY, *Convergence of Probability Measures*, Wiley, 1968.

[4] P. BILLINGSLEY, *Probability and Measure*, Wiley, 1986.

[5] R. M. BLUMENTHAL – R. K. GETOOR, *Markov processes and potential theory*, Pure and Applied Mathematics, Academic Press, New York-London, 1968.

[6] E. BOREL, *Leçons sur les fonctions discontinues*, Paris, 1898.

[7] N. BOURBAKI, *Eléments d'histoire des mathématiques*, Masson Editeur, Paris, 1984.

[8] C. CARATHÉODORY, *Vorlesung über reelle Funktionen*, Leipzig-Berlin, 1918.

[9] S. CERRAI, *Second order PDE's in finite and infinite dimension. A probabilistic approach*, Lecture Notes in Mathematics, 1762. Springer-Verlag, Berlin, 2001.

[10] Z. CIESIELSKI, *Hölder condition for realizations of Gaussian processes*, Trans. Amer. Math. Society **99** (1961), 403-413.

[11] PH. COURRÈGE, *Sur la forme integro-différentielle des operateurs de C_K^∞ dans C_0 satisfaisant au principe du maximum*, in "Sém. Théorie du Potentiel 1965/66", Exposé 2).

[12] P. J. DANIELL, *Integrals in an infinite number of dimensions*, Ann. of Math. (2) **20** (1918), 281-288.

[13] G. DA PRATO – J. ZABCZYK, *Stochastic equations in infinite dimensions*, Encyclopedia of Mathematics and its Applications **44**, Cambridge University Press, 1992.

[14] G. DA PRATO – J. ZABCZYK, *Ergodicity for Infinite Dimensional Systems*, Cambridge University Press, Cambridge, 1996.

[15] G. DA PRATO – J. ZABCZYK, *Second Order partial differential equations in Hilbert spaces*, London Mathematical Society Lecture Note Series, **293**, Cambridge University Press, Cambridge, 2002.

[16] E. B. DAVIS, *One-parameter Semigroups*, Academic Press, 1980.

[17] C. DELLACHERIE, *Un survol de la theorie de l'integral stochastique*, Stochastic Processes and Applications **10** (1980), 115-144.

[18] J. L. DOOB, *The law of large numbers for continuous stochastic processes*, Duke Math. J. **6** (1940), 290-306.

[19] J. L. DOOB, *Stochastic Processes*, John Wiley and Sons, 1953.

[20] E. B. DYNKIN, *Markov processes*, **I, II** Springer-Verlag, Berlin-Göttingen-Heidelberg, 1965.

[21] A. EINSTEIN, *Zur Theorie der Brownschen Bewegung*, Ann. Phys. IV **19** (1906), 371-381.

[22] K. D. ELWORTHY – X. M. LI, *Formulae for the Derivatives of Heat Semigroups*, J. Funct. Anal. **125** (1994), 252-286.

[23] K. ENGEL – R. NAGEL, *One-parameter Semigroups for Linear Evolution Equations*, Springer Graduate Texts in Mathematics **194**, 2000.

[24] S. N. ETHIER – T. G. KURTZ, *Markov processes. Characterization and convergence*, Wiley Series in Probability and Mathematical Statistics: "Probability and Mathematical Statistics". John Wiley Sons, Inc., New York, 1986.

[25] A. K. ERLANG, *The theory of probabilities and telephone conversations*, Nyt Tidsskrift für Matematik, B. **20** (1909), 33.

[26] W. FELLER, *An introduction to Probability Theory and Its Applications*, second edition **II** Wiley, 1971.

[27] M. FRÉCHET, *Sur l'intégral d'une fonctionnelle entendue à un ensemble abstrait*, Bull. Soc. Math. France **43** (1915).

[28] I. I. GIHMAN – A. V. SKOROHOD, *The Theory of Stochastic Processes*, **I, II, III**, Springer, 1979-1982.

[29] G. A. HUNT, *Markoff processes and potentials*, I, II, III, Illinois J. Math. **1** (1957), 44-93, 316-369; **2** (1958), 151-213.

[30] N. IKEDA – S. WATANABE, *Stochastic differential equations and diffusion processes*, North-Holland Mathematical Library, **24**, 1981.

[31] K. ITO, *On stochastic differential equations*, Mem. Amer. Math. Soc. **4** (1951), 1-51.

[32] J. JACOD – A. N. SHIRYAEV, *Limit Theorems for Stochastic Processes*, Springer Verlag, 1987.

[33] J. F. C. KINGMAN, *Poisson Processes*, Clarendon Press, Oxford, 1993.

[34] A. KOLMOGOROFF, *Grundbegriffe der Wahrscheinlichkeitsrechnung*, Erg. der Math., Bd 2, Berlin, Springer, 1933.

[35] A. KOLMOGOROFF, *Über die analytischen Methoden in der Wahrscheinlichkeitsrechnung*, Math. Ann. **104** (1931), 415-458.

[36] N. V. KRYLOV, *Introduction to the theory of Diffusion Processes*, American Mathematical Society, Providence, 1995.

[37] M. MÉTIVIER – G. PISTONE, *Une formule d'isométrie pour l'intégral stochastique d'évolution linéaire stochastiques*, Z. Wahrscheinlichkeitstheorie verw. Gebiete **33** (1975), 1-18.

[38] M. MÉTIVIER – J. PELLAUMAIL, *Stochastic Integration*, Academic Press, 1980.

[39] M. MÉTIVIER, *Semimartingales*, De Gruyter, 1982.

[40] M. MÉTIVIER, *Stochastic Partial Differential Equations in Infinite Dimensional Spaces*, Quaderni, Scuola Normale Superiore, 1988.

[41] H. LEBESGUE, *Integral, longeur, air*, Ann. Math. (3) **VII** (1902), 231-359.

[42] P. LÉVY, *Processus stochastiques et mouvement brownien*, Gauthier-Villars, Paris, 1948.

[43] F. LUNDBERG, *Zur Theorie def Rückversicherung*, Verhandl. Kongr. Versicherungsmath. Wien, 1909.

[44] P. MALLIAVIN, *Stochastic calculus of variation and hypo-elliptic operators*, Proc. Int. Symp. Stoch. Diff. Eqns, Kyoto, 1976; Kinokunigen-Wiley, 1978.

[45] P. A. MEYER, *Un cours sur les intégrales stochastiques*, LNiM **511** (1976), 245-400.

[46] D. MUMFORD, *The dawning of the age of stochasticity*, Mathematics, Frontiers and Perspectives, 2000.

[47] O. NIKODYM, *Sur une genéralisation des intégrales de M. J. Radon*, Fund. Math. **15** (1930), 131-179.

[48] D. NUALART, *The Malliavin Calculus and Related Topics*, Springer, 1995.

[49] K. B. OLDHAM – J. SPANIER, *The fractional calculus*, Academic Press, New York-London, 1974.

[50] R. E. A. C. PALEY – N. WIENER, *Fourier transforms in the complex domain*, Colloq. Publ. Amer. Math. Soc. (1934).

[51] YU. V. PROHOROV, *Convergence of random processes and limit theorems in probability*, Th. Prob. Appl. **1** (1956), 157-214.

[52] L. C. G. ROGERS – D. WILLIAMS, *Diffusions, Markov Processes and Martingales*, Cambridge University Press, **1-2** (2000).

[53] S. Saks, *Zarys Teorii Calki*, Warszawa, 1930. *Theory of the Integral*, 2nd ed. Warszawa, 1937, Monografie Matematyczne n. 7.

[54] K.-I. Sato, *Lévy processes and infinite divisible distributions*, Cambridge University Press, 1999.

[55] M. Sharpe, *General theory of Markov processes*, Pure and Applied Mathematics, Academic Press, Boston, MA, 1988.

[56] E. E. Slucky, *Alcuni proposizioni sulla teoria delle funzioni aleatorie*, Giorn. Ist. Ital. Attuari **8** (1937), 183-199.

[57] H. Steinhaus, *Les probabilités denombrales et leur rapport à la théorie de la mesure*, Fund. Math. **4** (1922), 286-310.

[58] J. C. Oxtoby – S. Ulam, *On the existence of a measure invariant under a transformation*, Ann. Math. **2** (1939), 560-566.

[59] D. W. Stroock – S. R. S. Varadhan, *Multidimensional diffusion processes*, Springer, 1979.

[60] J. Ville, *Critique de la notion de collectif*, Paris, 1939.

[61] N. Wiener, *Differential space*, J. Math. Phys., Math. Inst. Tech. **2** (1923), 131-174.

[62] J. Zabczyk, *The fractional calculus and stochastic evolution equations*, In: "Barcelona Seminar on Stochastic Analysis", D. Nualart – M. Sans-Solé (eds.), Birkhauser, Basel, Boston, Berlin, 1993, pp. 222-234.

[63] J. Zabczyk, *Parabolic equations on Hilbert spaces*, Lecture Notes in Mathematics **1715** (1999), 117-213.

[64] J. Zabczyk, *Leverhulme Lectures*, Preprint 8 (2001), Department of Mathematics, University of Warwick

PUBBLICAZIONI DELLA CLASSE DI SCIENZE
DELLA SCUOLA NORMALE SUPERIORE

QUADERNI

1. DE GIORGI E., COLOMBINI F., PICCININI L.C.: *Frontiere orientate di misura minima e questioni collegate.*
2. MIRANDA C.: *Su alcuni problemi di geometria differenziale in grande per gli ovaloidi.*
3. PRODI G., AMBROSETTI A.: *Analisi non lineare.*
4. MIRANDA C.: *Problemi in analisi funzionale* (ristampa).
5. TODOROV I.T., MINTCHEV M., PETKOVA V.B.: *Conformal Invariance in Quantum Field Theory.*
6. ANDREOTTI A., NACINOVICH M.: *Analytic Convexity and the Principle of Phragmén-Lindelöf.*
7. CAMPANATO S.: *Sistemi ellittici in forma divergenza. Regolarità all'interno.*
8. TOPICS IN FUNCTIONAL ANALYSIS: *Contributors:* F. STROCCHI, E. ZARANTONELLO, E. DE GIORGI, G. DAL MASO, L. MODICA.
9. LETTA G.: *Martingales et intégration stochastique.*
10. OLD AND NEW PROBLEMS IN FUNDAMENTAL PHYSICS: *Meeting in honour of* GIAN CARLO WICK.
11. INTERACTION OF RADIATION WITH MATTER: *A Volume in honour of* ADRIANO GOZZINI.
12. MÉTIVIER M.: *Stochastic Partial Differential Equations in Infinite Dimensional Spaces.*
13. SYMMETRY IN NATURE: *A Volume in honour of* LUIGI A. RADICATI DI BROZOLO.
14. NONLINEAR ANALYSIS: *A Tribute in honour of* GIOVANNI PRODI.
15. LAURENT-THIÉBAUT C., LEITERER J.: *Andreotti-Grauert Theory on Real Hypersurfaces.*
16. ZABCZYK J.: *Chance and Decision. Stochastic Control in Discrete Time.*
17. EKELAND I.: *Exterior Differential Calculus and Applications to Economic Theory.*
18. ELECTRONS AND PHOTONS IN SOLIDS: *A Volume in honour of* FRANCO BASSANI.
19. ZABCZYK J.: *Topics in Stochastic Processes.*

CATTEDRA GALILEIANA

1. LIONS P.L.: *On Euler Equations and Statistical Physics.*
2. BJÖRK T.: *A Geometric View of the Term Structure of Interest Rates.*
3. DELBAEN F.: *Coherent Risk Measures.*

LEZIONI LAGRANGE

1. VOISIN C.: *Variations of Hodge Structure of Calabi-Yau Threefolds.*

LEZIONI FERMIANE

1. THOM R.: *Modèles mathématiques de la morphogénèse.*
2. AGMON S.: *Spectral Properties of Schrödinger Operators and Scattering Theory.*
3. ATIYAH M.F.: *Geometry of Yang-Mills Fields.*
4. KAC M.: *Integration in Function Spaces and Some of Its Applications.*
5. MOSER J.: *Integrable Hamiltonian Systems and Spectral Theory.*
6. KATO T.: *Abstract Differential Equations and Nonlinear Mixed Problems.*
7. FLEMING W.H.: *Controlled Markov Processes and Viscosity Solution of Nonlinear Evolution Equations.*
8. ARNOLD V.I.: *The Theory of Singularities and Its Applications.*
9. OSTRIKER J.P.: *Development of Larger-Scale Structure in the Universe.*
10. NOVIKOV S.P.: *Solitons and Geometry.*
11. CAFFARELLI L.A.: *The Obstacle Problem.*
12. CHEEGER J.: *Degeneration of Riemannian metrics under Ricci curvature bounds.*

PUBBLICAZIONI DEL CENTRO DI RICERCA MATEMATICA ENNIO DE GIORGI

PROCEEDINGS

1. *DYNAMICAL SYSTEMS.* Part I: *Hamiltonian Systems and Celestial Mechanics.*
2. *DYNAMICAL SYSTEMS.* Part II: *Topological, Geometrical and Ergodic Properties of Dynamics.*

PUBLIC LECTURES

1. *MATEMATICA, CULTURA E SOCIETÀ 2003.*

ALTRE PUBBLICAZIONI

Proceedings of the Symposium on FRONTIER PROBLEMS IN HIGH ENERGY PHYSICS Pisa, June 1976

Proceedings of International Conferences on SEVERAL COMPLEX VARIABLES, Cortona, June 1976 and July 1977

Raccolta degli scritti dedicati a JEAN LERAY apparsi sugli Annali della Scuola Normale Superiore di Pisa

Raccolta degli scritti dedicati a HANS LEWY apparsi sugli Annali della Scuola Normale Superiore di Pisa

Indice degli articoli apparsi nelle Serie I, II e III degli Annali della Scuola Normale Superiore di Pisa (dal 1871 al 1973)

Indice degli articoli apparsi nella Serie IV degli Annali della Scuola Normale Superiore di Pisa (dal 1974 al 1990)

ANDREOTTI A.: *SELECTA vol. I, Geometria algebrica.*

ANDREOTTI A.: *SELECTA vol. II, Analisi complessa, Tomo I e II.*

ANDREOTTI A.: *SELECTA vol. III, Complessi di operatori differenziali.*

Fotocomposizione "CompoMat" Loc. Braccone, 02040 Configni (RI), Italy
Finito di stampare per conto della "CompoMat" dalla Nuova Grafica 86 nel luglio 2004